STUDY GUIDE TO ACCOMPANY

PHYSICS

PHYLLIS J. FLEMING
WELLESLEY COLLEGE

ADDISON-WESLEY PUBLISHING COMPANY
Reading, Massachusetts • Menlo Park, California
London • Amsterdam • Don Mills, Ontario • Sydney

Reproduced by Addison-Wesley from camera-ready copy prepared by the author.

ISBN 0-201-02474-8
ABCDEFGHIJ-AL-7987

PREFACE

The purpose of the Study Guide is to help you do well in the course. The sections that correspond to chapters in the text include: (1) Objectives, (2) Trouble Spots (3) Illustrative Examples, and (4) Problems. Through the Objectives, I try to stress the concepts in each chapter that you should master. The section on Trouble Spots tries to warn you about common errors. The level of difficulty for the illustrative examples is greater than that of the problems. I strongly recommend that you read the statement of the problems in the illustrative examples and try to solve them before reading the answers. If I had a nickel for every student, who has said to me, "I understand your lectures and the text, but I can't do the problems," I would be very wealthy. The truth is you do not understand the material unless you can do the problems.

After two or three chapters, you will find sample quizzes. The answers to the problems and the sample quizzes are at the end of the Study Guide. Again try the problems and the sample quizzes before you look up the answers. You will need to refer to the Appendix of the text for trigonometric functions.

The Introduction is a quick review of the mathematics necessary in the study of physics. Many of you will find it superfluous. For those of you, who have been "away" from the use of mathematics for a while, I strongly recommend that you read it and try to do the problems. Too frequently, I have seen students, who have a real feeling for physics, slowed down by their lack of facility with mathematics. Don't let that happen to you.

Unfortunately the way of the world is usually to have a time limit on quizzes. You can save enormous amount of time by first solving a problem algebraically. I'll give an example solved in two different ways. Follow through the examples, do the arithmetic, and time yourself for both approaches.

You are given the following relations:

$$1/2\ mv^2 = qV_{ab} \qquad (1)$$

$$Bqv = mv^2/r \qquad (2)$$

The meaning of the symbols are as follows: m is the mass of the object, v its speed, q its charge, r the radius of its circular path in a magnetic field B. V_{ab} is the potential difference through which it is accelerated.

Let $m = 10^{-10}$ kg, $q = 10^{-6}$ C, $V_{ab} = 2 \times 10^2$ J C^{-1}, and $B = 10^{-1}$ N $(A\text{-}m)^{-1}$) Find r.

Method 1.

$$1/2\ mv^2 = qV_{ab}$$

$$v = \sqrt{2qV_{ab}/m}$$

$$v = \frac{2 \times 10^{-6}\ C \times 2 \times 10^2\ J\ C^{-1}}{10^{-10}\ kg}$$

$$v = \sqrt{4 \times 10^6\ J/kg}$$

$$= 2 \times 10^3\ m\ s^{-1}$$

$$Bqv = mv^2/r$$

$$r = mv/Bq$$

$$r = \frac{10^{-10}\ kg \times 2 \times 10^3\ m\ s^{-1}}{10^{-1}\ N(A\text{-}m)^{-1} \times 10^{-6}\ C}$$

$$= 2\ m$$

Method 2.

$$v = \sqrt{2qV_{ab}/m}$$

$$r = mv/Bq = (m/Bq)\sqrt{2qV_{ab}/m}$$

$$r = (1/B)\sqrt{2V_{ab}m/q}$$

$$r = \frac{1}{10^{-1}\ N(A\text{-}m)^{-1}} \frac{2 \times 2 \times 10^2\ J\ C^{-1} \times 10^{-10} kg}{10^{-6}\ C}$$

$$= \frac{1}{10^{-1}\ N\ (A\text{-}m)^{-1}} \sqrt{4 \times 10^{-2}\ kg^2\ m^2\ C^{-2}\ s^{-2}}$$

$$= 2\ m$$

I'm certain that in the above example you forgot about the units and just carried on. Nevertheless, it helps to check units. It will save you some very embarrassing moments.

To follow through a check of units, you need to know that

$$N = kg\ m\ s^{-2}$$

$$J = N\text{-}m = kg\ m^2\ s^{-2}$$

$$A = C\ s^{-1}$$

For Method 1,

$$J/kg = kg\ m^2\ s^{-2}/kg = m^2\ s^{-2}$$

and

$$kg\ m\ s^{-1}/N(A\text{-}m)^{-1}C = kg\ m^2\ C\ s^{-2}/kg\ m\ s^{-2}C$$

$$= m$$

For Method 2,

$$\frac{1}{N(A\text{-}m^{-1})}\sqrt{J\ C^{-1}\ kg/C} = \frac{C\ s^{-1}m}{kg\ m\ s^{-2}}\sqrt{kg^2\ m^2\ s^{-2}/C^2}$$

$$= m$$

Good luck in the course. I want you to do well and enjoy it.

Phyllis J. Fleming
Wellesley College

CONTENTS

INTRODUCTION. A Review of Mathematics

This chapter is a quick review of the mathematics necessary in the study of physics. It is important that you not allow a deficiency in mathematical skills deter you from doing well in the course.

First we review the rules for exponents:

To multiply, add exponents: $x^a \cdot x^b = x^{a+b}$

To divide, subtract exponents:

$$x^a \div x^b = x^{a-b}$$

To raise to the nth power, multiply exponents by n: $(x^a)^n = x^{na}$

To take the nth root, divide exponent by n:

$$\sqrt[n]{x^a} = x^{a/n}$$

Illustrative Examples

1. $6 \times 10^{14} \times 5 \times 10^{-7} = 30 \times 10^{7} = 3 \times 10^{8}$

2. $6.63 \times 10^{-34}/10^{-10} = 6.63 \times 10^{-24}$

3. $(2 \times 10^3)^2 = 4 \times 10^6$

4. $(0.6)^2 = 0.36$

5. $\sqrt{1 - (0.6)^2} = \sqrt{0.64} = 0.8$

6. $1/5 \times 10^{-5} = 0.2 \times 10^{-5} = 2 \times 10^{-6}$

7. $\dfrac{10^2 + 10^5}{10^2} = 10^0 + 10^3 = 1 + 1000 = 1001$

8. $(10^{-8} \div 10^4) \times 10^{-6} = 10^{-18}$

Exponents for units are handled the same way as they are for numbers. For example, the unit of displacement s is meter (m) and the unit of time t is second (s). Average velocity is defined as the displacement divided by the time to make the displacement. The unit of velocity is

$$\frac{(m)}{(s)} = m\ s^{-1}$$

Acceleration is the change in velocity divided by the time to make the change. The unit of acceleration is

$$\frac{(m/s)}{(s)} = m/s^2 = m\ s^{-2}$$

Solutions to $ax^2 + bx + c = 0$ are

$$x = \frac{-b \pm \sqrt{b^2 - 4ac}}{2a}$$

Illustrative Examples

1. Find the roots of $2x^2 - 3x - 5 = 0$.

$$x = \frac{3 \pm \sqrt{9 + 40}}{4}$$

$x = 2.5$ and -1

2. Find the roots of $x^2 - 10 = 4x^2 + 3x - 46$

This equation reduces to $3x^2 + 3x - 36 = 0$ or $x^2 + x - 12 = 0$.

$$x = \frac{-1 \pm \sqrt{1 + 48}}{2}$$

$x = 3$ or -4

When you have two equations and two unknowns, you should solve for one unknown in one of the equations and then plug this result into the other equation.

Illustrative Examples

1. Given $4x + 3y = 18$ (1) and $2x - y = 4$ (2), solve for x and y.

From Eq. (1), $x = (18 - 3y)/4$. Substituting this value of x in Eq. (2), we find

$$2(18-3y)/4 - y = 4$$

$$(18 - 3y)/2 = 4 + y$$

$$18 - 3y = 8 + 2y$$

$$10 = 5y$$

$$y = 2$$

Substituting the value of y into Eq. (2), we find

$$2x - 2 = 4$$

$$x = 3$$

Alternatively, you can solve for x and y in the following way:

Multiply Eq. 2 by 2: $4x - 2y = 8$ (3)

Write down Eq. 1: $4x + 3y = 18$ (1)

Then Eq. (1) - Eq. (3): $5y = 10$

$$y = 2$$

Now substitute the value of y back into either Eq. 1 or 2 to find x.

2. Given $3x + y + 1 = 0$ (1) and $y - 2x = 9$ (2), find x and y.

From (2), $y = 9 + 2x$

Substitute this into (1): $3x + (9 + 2x) + 1 = 0$

$$5x = -10$$

$$x = -2$$

$$y = 9 + 2x = 9 + 2(-2) = 5$$

3. Given $hf = hf' + 1/12\ m_oc^2$ and $hf = -hf' + 5\ m_oc^2/12$, find f.

$$hf - hf' = 1/12\ m_oc^2 \qquad (1)$$

$$hf + hf' = 5/12\ m_oc^2 \qquad (2)$$

(1) + (2): $2hf = 6/12\ m_oc^2$

$$f = 1/4\ m_oc^2/h$$

The sine of an angle in a right triangle is defined as the side opposite divided by the hypotenuse. The cosine is the side adjacent divided by the hypotenuse. The tangent is the side opposite divided by the side adjacent.

Illustrative Examples

1. Find the sine, cosine, and the tangent of 53° (Fig. I.1).

(a) $\sin 53^o = 4/5 = 0.8$

(b) $\cos 53^o = 3/5 = 0.6$

(c) $\tan 53^o = 4/3$

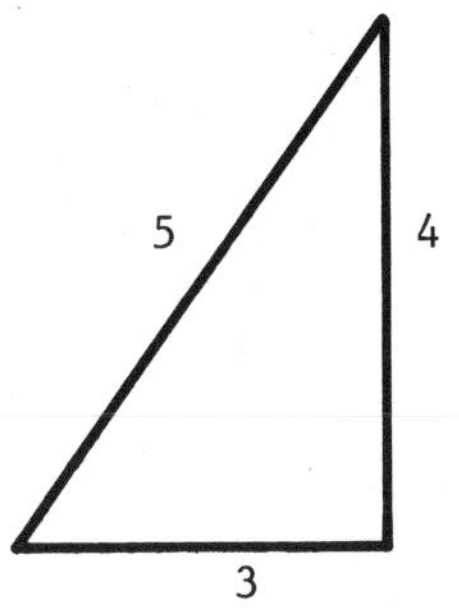

Fig. I.1

2. Find the sine and cosine of 62° (Fig. I.2).

Using the Pythagorean Theorem, we find

hypotenuse =

$$\sqrt{(8)^2 + (15)^2}$$

$$= \sqrt{64 + 225}$$

$$= \sqrt{289} = 17$$

$\sin 62^o = 15/17$

$\cos 62^o = 8/17$

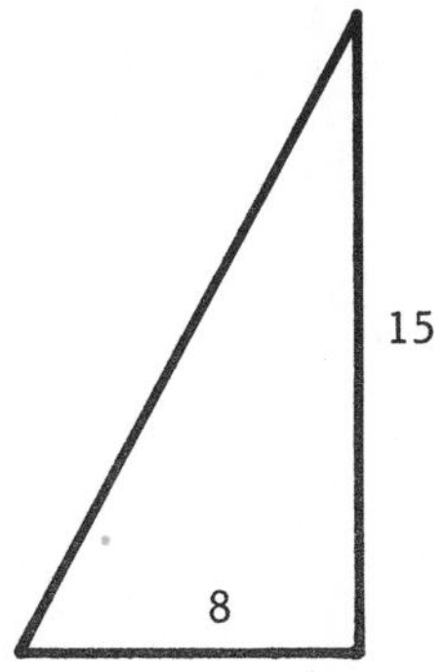

Fig. I.2

An angle in radians equals the length of arc it subtends divided by the radius. 1 radian = $180^{\circ}/\pi$.

Illustrative Examples

1. Express $\pi/8$ radians in degrees.

$$1 \text{ radian} = 180^{\circ}/\pi$$

$$\pi/8 \text{ radian} = \pi/8 \times 180^{\circ}/\pi = 22.5^{\circ}$$

2. Express 30° in radians.

$$1^{\circ} = \pi \text{ radian}/180$$

$$30^{\circ} = 30\,\pi \text{ radian}/180 = \pi/6$$

3. Express 45° in radians.

$$45^{\circ} = 45\,\pi \text{ radian}/180 = \pi/4$$

The following are useful trigonometric identities:

$$\sin^2 x + \cos^2 x = 1$$

$$\sin(x \pm y) = \sin x \cos y \pm \cos x \sin y$$

$$\cos(x \pm y) = \cos x \cos y \mp \sin x \sin y$$

Illustrative Examples

1. Given that $\sin 32^{\circ} = 0.53$, evaluate the following: (a) $\cos 32^{\circ}$, (b) $\sin 58^{\circ}$, (c) $\cos 238^{\circ}$.

(a) $\sin^2 32^{\circ} + \cos^2 32^{\circ} = 1$

$$\cos^2 32^{\circ} = 1 - (0.53)^2$$

$$= 1 - 0.28 = 0.72$$

$$\cos 32^{\circ} = 0.85$$

(b) $58^{\circ} = 90^{\circ} - 32^{\circ}$

$$\sin(90^{\circ} - 32^{\circ}) = \sin 90^{\circ}\cos 32^{\circ} + \cos 90^{\circ}\sin 32^{\circ}$$

$$= (1)(0.85) + (0)(0.53) = 0.85$$

(c) $238^{\circ} = 270^{\circ} - 32^{\circ}$

$$\cos(270^{\circ} - 32^{\circ}) = \cos 270 \cos 32 + \sin 270 \sin 32$$

$$= (0)(0.85) + (-1)(0.53)$$

$$= -0.53$$

You will need to remind yourself of the following information from geometry:

The area of a right triangle (Fig. I.3) is 1/2 bh.

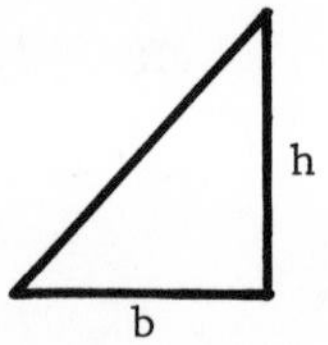

Fig. I.3

The area of a oblique triangle (Fig. I.4) is 1/2 bh.

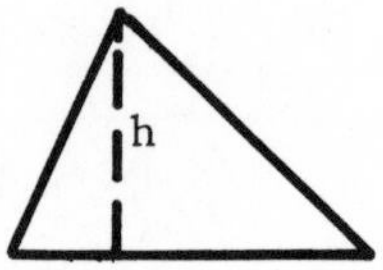

Fig. I.4

The area of a square of sides a is a^2.

The area of a rectangle of sides a and b is ab.

The circumference of a circle of radius r is $2\pi r$.

The area of a circle is πr^2.

The surface area of a sphere of radius r is $4\pi r^2$.

The volume of a sphere is $4/3\ \pi r^3$.

The surface area of a right cylinder of radius r and length ℓ is $2\pi r\ell$.

The volume of a right cylinder is $\pi r^2 \ell$.

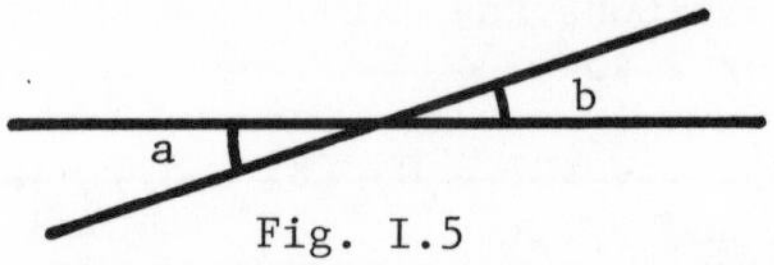

Fig. I.5

The vertical angles a and b in Fig. I.5 are equal.

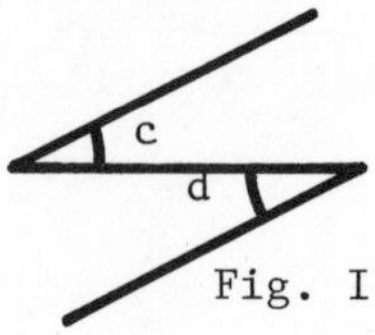

Fig. I.6

Angles c and d in Fig. I.6 are equal because their sides are parallel.

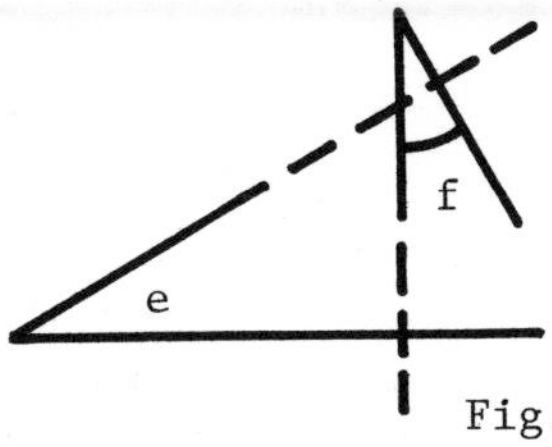

Fig. I.7

Angles e and f in Fig. I.7 are equal because their sides are mutually perpendicular.

Illustrative Examples

1. The area of a right triangle with base b = 10 cm is 30 cm^2. What is the height of the triangle?

   Since A = 1/2 bh, h = 2A/b = 60 cm^2/10 cm or h = 6 cm.

2. A square has sides a. One side of a rectangle is a. What is the other side b of the rectangle if its area is four times the area of the square?

   Area of square = a^2; Area of rectangle = ab
   Thus ab = $4a^2$ or b = 4a.

3. The circumference of a circle is π^2 cm. What is its area?

   Circumference = $2\pi r = \pi^2$ cm or $r = \pi/2$ cm

   Its area = $\pi r^2 = \pi(\pi/2 \text{ cm})^2 = \pi^3/4 \text{ cm}^2$

4. A sphere and a cylinder both of radius r have the same surface area. What is the length of the cylinder?

   Surface area of sphere = $4\pi r^2$

   Surface area of cylinder = $2\pi r\ell$

   $2\pi r\ell = 4\pi r^2$ or $\ell = 2r$

5. You double the radius of a sphere. What happens to its volume?

   Initial volume = $4/3\ \pi r^3$

   Final volume $= 4/3\ \pi(2r)^3$

   $= 8 \times 4/3\ \pi r^3$

   = 8 x Initial volume

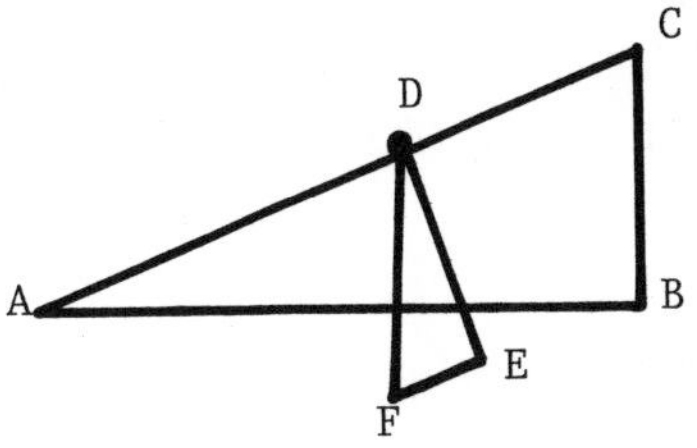

Fig. I.8

6. In Fig. I.8, ∠CAB is equal to what angle in Δ DEF?

   DF is perpendicular to AB and ED is perpendicular to AC so ∠EFD = ∠CAB.

The following are special cases of the binomial expansion:

$$\frac{1}{1 \pm x} = 1 \mp x + x^2 \mp x^3 \ldots.$$

$$\frac{1}{(1 \pm x)^2} = 1 \mp 2x + 3x^2 \mp 4x^3 \ldots.$$

$$\sqrt{1 \pm x} = 1 \pm 1/2\ x - 1/8\ x^2 \pm 1/16\ x^3$$

Illustrative Examples

1. Estimate $1/\sqrt{4 + 0.16}$ to two significant figures.

$$\frac{1}{\sqrt{4 + 0.16}} = \frac{1}{2\sqrt{1 + 0.04}}$$

$$= 1/2(1 - 0.02 + 0.0006)$$

$$= 0.49$$

If we had included the 0.0006 in our calculation, we would have had more than two significant figures.

2. Expand $1/\sqrt{1 - v^2/c^2} - 1$ including no power of v higher than 2.

$$\frac{1}{\sqrt{1 - v^2/c^2}} = (1 + 1/2\ v^2/c^2)$$

$$\frac{1}{\sqrt{1 - v^2/c^2}} - 1 = (1 + 1/2\ v^2/c^2 - 1)$$

$$= 1/2\ v^2/c^2$$

Problems

1. One fourth of 10^{-4} is ______________

2. $(2a + 4b)/2 =$ ________________

3. $(x + a)(x - a) =$ ____________

4. $1 \div 10^3 =$ ____________

5. $(3 \times 10^8)^2 =$ ____________

6. $(8)^{1/3} =$ ____________

7. The reciprocal of 10^{-3} is ___________

8. $\sqrt{2} \times \sqrt{2} =$ ________________

9. $\frac{1}{A} + \frac{1}{B} =$ ________________

10. The reciprocal of $(1/A) + (1/B) =$ _________

11. The reciprocal of $\frac{1}{3} + \frac{1}{4} + \frac{1}{12}$ is ______________

12. The square root of $(9)^4$ is _____________

13. The roots of $x^2 - 4x - 5$ are ___ & ___

14. The roots of $2x^2 - 28 = 10x$ are ___&___

15. Given $2x + y = 3$ and $6x - 2y + 1 = 0$, you find that x = _____and y = _______.

16. Given $3x - 2y = 1$ and $4y - 5x = 1$, you find that x = ____ and y = _____.

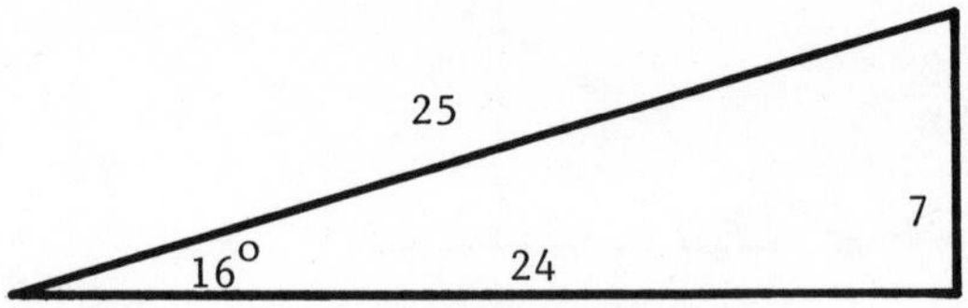

Fig. I.9

17. The tangent of 16° (Fig. I.9) is ______

18. The sine of 74° is ________

19. The cosine of 106° is _______

20. 67.5° in radians is ________

21. The cosine of $\pi/3$ is _________

22. The height of a right triangle is kx and its base is x. The area of the triangle is ____________.

23. A cylindrical glass has a radius of 3 cm and a height of 10 cm. The density d of water is 1 g cm^{-3}. Given that mass = density x volume, the mass of water contained in such a glass is ___________

24. A bubble has a diameter of 4 cm. Its surface area is ___________

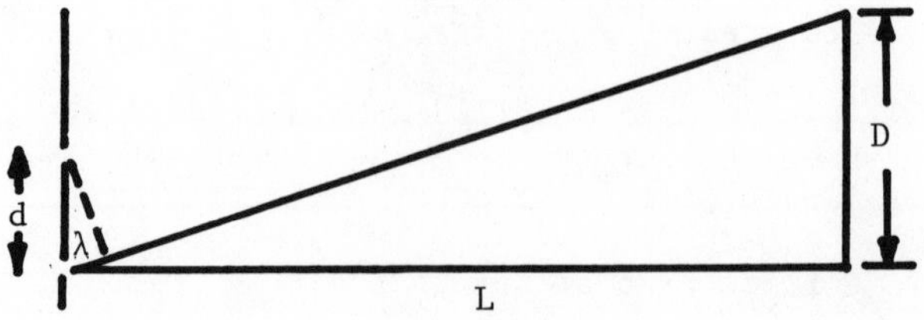

Fig. I.10

25. In Fig. I.10, d = 0.01 cm, $\lambda = 5 \times 10^{-5}$ cm, and $\sqrt{L^2 + D^2} = 100$ cm. D = _______

CHAPTER 1. The Program of Physics

Objectives

You should:

1. recognize that an object can be subjected to individual forces, but remain at rest or move with a constant speed in a straight line

2. learn that a net force is necessary to produce an object's change in motion

3. differentiate between Newton's and Aristotle's concept of the result of a net force

    Newton

    (a) Takes an object from rest to motion

    (b) Takes an object from motion to rest

    (c) Changes the speed of an object

    (d) Changes the direction of motion

    Aristotle

    (a) Takes an object from rest to motion

    (b) Takes an object from motion to rest

    (c) Changes the speed of an object

    (d) Changes the direction of motion

    (e) Keeps an object moving with a constant speed in a straight line

4. be able to draw diagrams of observers moving relative to each other with a constant speed and show how they explain the report of the other observer about the results of an experiment.

Trouble Spots

When you find the net force acting on an object, look at all the individual force's acting on it. For example, when you hold a book, there is a gravitational force downward and your force upward. If the book is at rest or moving with a constant speed in a straight line, the force upward cancels the force downward, and the net force is a summation of forces and a sum is singular, it is incorrect to say net force_s_.

A net force is the cause of a change in motion. The effect of a net force is a change in the speed of an object or a change in its direction.

Because it is impossible to do an experiment to detect whether you are at rest or moving with a constant speed in a straight line, motion is relative. A person on a train can maintain that he is at rest and the platform moves to the left. Someone on the platform can say that she is at rest and that the train moves to the right.

Illustrative Examples

1. A person paddles a canoe in a straight line with a constant speed. What is the net force acting on the canoe.

   Since the canoe moves with a constant speed in a straight line, the net force acting on the canoe must be zero. The person's effort is neutralized by the frictional force of the water.

2. An object experiences no net force. What can you say about the magnitude of the speed? What can you say about the direction of motion of the object?

   You cannot tell how great the speed of the object is. It could be 5 m s^{-1}, 10 m s^{-1}, and so forth. The one thing you do know is that the magnitude of the speed remains the same. You also know that the direction of the object does not change because no net force acts on it.

3. Two children at the opposite ends of the car of a train, moving with a constant speed in a straight line, roll balls toward the center of the train with the same speed. Will the balls arrive at the center of the car (a) from the point of view of the children (b) from the point of view of an observer standing on the platform at the same time?

   (a) The children consider that they are at rest. Since they give the balls equal speeds, they say that the balls arrive at the center of the car at the same time.

   (b) The observer on the platform must agree with the children that the balls will reach the center of the train at the same time or the principle of relativity would be violated. While this may seem a strange result to you, remember that from the point of view of the observer on the platform, the center point moves in the same direction as the train. Let us say that the train moves to the right. Then the ball rolled to the right has its given speed <u>and</u> the speed of the train to the right, while the ball rolled to the left has its given speed minus the speed of the train. Since the center point moves to the right, the ball rolled to the right has to move a greater distance than the one rolled to the left. The former can do this because it has a greater net speed than the ball that moves to the left.

<u>Problems</u>

1. An object is at rest.

   (a) No forces act on the object

   (b) No net force acts on the object

2. An object moves with a constant speed of 5 miles per hour in a straight line. No net force acts on the object. The object

   (a) will slow down but never come to rest

   (b) continue at a speed of 5 miles per hour

   (c) slows down and comes to rest

3. A person holds a book in her hand. To hold two books, she must

   (a) exert the same force

   (b) exert a greater force

4. While riding in a train, that moves with a constant speed in a straight line, a girl tosses a ball into the air. The ball will fall

   (a) behind her

   (b) in front of her

   (c) into her hands

5. A box is given a shove along a floor and comes to rest. According to a follower of Newton, the box comes to rest because

   (a) no net force acts on the box

   (b) a net force acts on the box

6. In Question 5, a follower of Aristotle says the box comes to rest because

   (a) no net force acts on the box

   (b) a net force acts on the box

7. A man on a train, moving with a constant speed in a straight line, drops a ball. There are observers on the train and on the platform.

   (a) The observer on the train says the ball will land at the man's feet, but the observer on the platform says it will fall behind the man's feet

   (b) Both observers say the ball will land at the man's feet

   (c) The observer on the train and the observer on the platform say the ball will fall behind the man's feet

8. A toy train moves around a circular track with a constant speed.

   (a) No net force acts on the train

   (b) A net force acts on the train

9. A pilot wishes to drop hay to stranded cattle. He should drop the hay

   (a) when the plane is directly over them

   (b) before the plane is directly over them

   (c) when the plane has passed them

10. A person on a train, moving to the right with a constant speed, drops a ball.

   (a) An observer on the platform says the ball falls behind the person's feet

   (b) An observer on the platform says the ball falls in a parabolic path but lands at the person's feet because the ball continues forward with the horizontal speed of the train.

11. An observer on the train of Problem 10 explains the report of the platform observer in the following way:

   (a) I am at rest and the platform moves to the left.

   (b) I know that I am moving to the right because I can detect my motion by doing an experiment.

CHAPTER 2. Language of Motion

Objectives

You should:

1. learn to differentiate between average velocity and instantaneous velocity

2. recognize the difference between velocity and acceleration

3. see that an object can experience an acceleration and at the same time have an instantaneous velocity of zero

4. master vector addition and subtraction

5. identify the slope of position versus time and slope of velocity versus time at any instant as the instantaneous velocity and instantaneous acceleration, respectively.

6. be able to show that the area under the velocity versus time curve and the area under the acceleration versus time curve as the distance moved by the object and its change in velocity, respectively

7. differentiate between motion with constant velocity and constant acceleration and learn how to find the displacement at any time for the two cases.

8. know the principle of superposition and be able to apply it to two-dimensional motion

Trouble Spots

Do not confuse average velocity with instantaneous velocity. When an object is dropped, it has an acceleration of approximately 10 m s^{-2}. If it is initially at rest, after 1 second it has an instantaneous velocity of 10 m s^{-1} because its velocity increases by 10 m s^{-1} in one second. In 1 second it moves a distance

$$s = v_o t + 1/2\, at^2$$

$$= (0)(1\ s) + 1/2 \times 10\ m\ s^{-2} \times 1\ s^2$$

$$= 5\ m$$

Thus during the interval of 1 second, it has an average velocity

$$\bar{v} = s/t = 5\ m/1\ s = 5\ m\ s^{-1}$$

While the velocity at t = 1 s equals 10 m s^{-1}, the average velocity during one second is 5 m s^{-1}.

Average velocity is the distance moved divided by the time to move that distance. The unit of velocity is m s^{-1}. Average acceleration is the change in velocity divided by the time to change the velocity. The unit of acceleration is m s^{-2}.

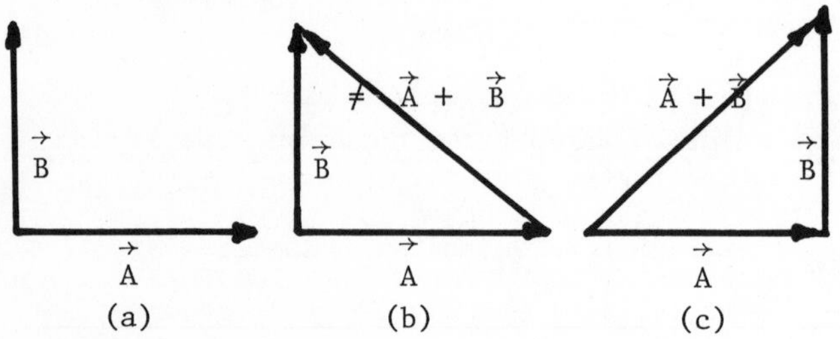

Fig. 2.1

Suppose you are asked to add vectors $\vec{A}$ and $\vec{B}$, shown in Fig. 2.1(a). An incorrect vector addition is shown in Fig. 2.1(b). In Fig. 2.1(c), the tail of $\vec{B}$ is moved to the head of $\vec{A}$ and the resultant goes from the tail of $\vec{A}$ to the head of $\vec{B}$.

In treating problems in two-dimensions, separate the problem into two one-dimensional problems. If the object has an initial velocity at an angle with the horizontal, you cannot use that velocity to determine the distance moved in a horizontal or vertical direction. Remember in projectile motion, there is no acceleration in a horizontal direction and the time involved for both the horizontal and vertical motion is the same.

Illustrative Examples

1. An object moves with a constant velocity of 5 m s^{-1} for 2 s and then accelerates with a constant acceleration of 5 m s^{-2} for 2 s. The motion is in a straight line. (a) Sketch a graph of the velocity as a function of time. (b) Find the distance moved in the total time of 4 s.

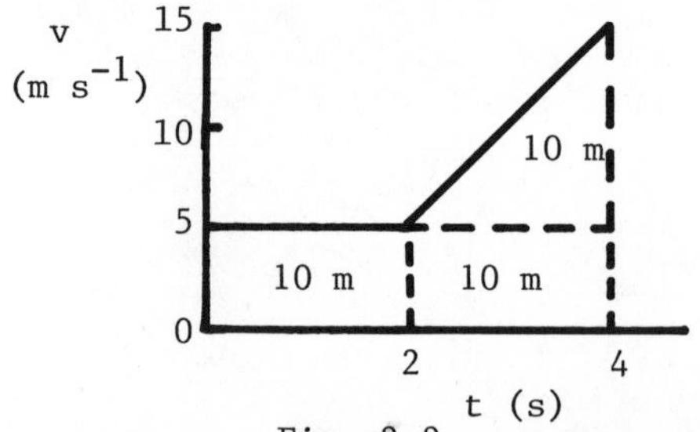

Fig. 2.2

(a) A plot of velocity versus time is shown in Fig. 2.2. Because the motion for the first two seconds is different from that of the last two seconds, the motion should be divided into two parts. From t = 0 to t = 2 s, the acceleration of the object is zero and it moves with a constant velocity. The graph from t = 0 to t = 2 s is a line parallel to the time-axis at v = 5 m s^{-1}.

In the next two seconds, the object has an acceleration of 5 m s^{-2} starting with an initial velocity at t = 2 s of 5 m s^{-1}. Some care must be taken in using the expression

$$v = v_o + at.$$

You want to find the instantaneous velocity at t = 4 s, but the acceleration takes place for only 2 s. At t = 2 s, $v_o = 5\ m\ s^{-1}$ and the acceleration takes place for 2 s. Thus, at t = 4 s

$$v = 5\ m\ s^{-1} + (5\ m\ s^{-2})(2\ s)$$

$$= 15\ m\ s^{-1}.$$

(b) The easiest way to find the distance moved in 4 s is to find the area under the velocity versus time curve of Fig. 2.2. The total distance moved, as shown in Fig. 2.2, if you add up the areas is 30 m.

If you use the formulas for displacement be careful to separate the motion into two parts from (1) t = 0 to t = 2 s and (2) from t = 2 s to t = 4 s. From t = 0 to t = 2 s, the object moves

$$s = v_o t + 1/2\ at^2$$

$$= (5\ m\ s^{-1})(2\ s) + 1/2\ (0)(4\ s^2)$$

$$= 10\ m$$

From t = 2 s to t = 4 s, the object moves

$$s = (5\ m\ s^{-1})(2\ s) + 1/2(5\ m\ s^{-2})(4\ s^2)$$

$$= 20\ m$$

Again, the total distance moved in 4 s is 30 m.

2. A person walks 17 mph for one hour at an angle of 62° north of east and then walks at a speed of 5 mph for two hours at an angle of 37° north of west. What is her average velocity (direction and magnitude)?

First you should find the distances moved. In the first hour, she moves

17 mph x 1 hr = 17 mi, 62° north of east

In the next two hours, she moves

5 mph x 2 hr = 10 mi, 37° north of west

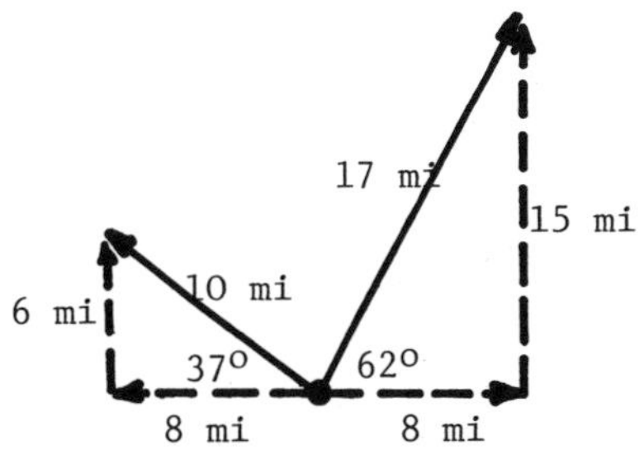

Fig. 2.3

The individual displacements and their components are shown in Fig. 2.3. The 17 mi displacement has a component of 8 mi to the east and 15 mi to the north. The 10 mi displacement has a component of 8 mi to the west and 6 mi to the north. Since the components to the east and west cancel, the resultant displacement equals

(6 + 15) mi = 21 m, to the north

The average velocity equals

21 mi/3 hr = 7 mph to the north.

3. A rock is thrown from the top of a building 24 m above the ground at a speed of $25\ m\ s^{-1}$ at an angle of 16° with the horizontal. Find (a) the time the rock is in the air, (b) the horizontal distance traveled by the rock, and (c) the horizontal and vertical velocities of the rock when it strikes the ground. Take g = $10\ m\ s^{-2}$.

First separate the problem into horizontal and vertical components and write down what you know. Take the upward direction to be positive. The symbols used below are illustrated in Fig. 2.4.

horizontal

$(v_o)_h = v_o \cos 16^o = (25 \text{ m s}^{-1})(24/25)$

$= 24 \text{ m s}^{-1}$

$a_h = 0$

$t = ?$

$s_h = (v_o)_h\, t$

$v_h = (v_o)_h$

vertical

$(v_o)_v = v_o \sin 16^o = (25 \text{ m s}^{-1})(7/25)$

$= 7 \text{ m s}^{-1}$

$a_v = -\ 10 \text{ m s}^{-2}$

$t = ?$

$s = -24 \text{ m} = (v_o)_v t + 1/2\ a_v\, t^2$

$v_v = (v_o)_v + a_v t$

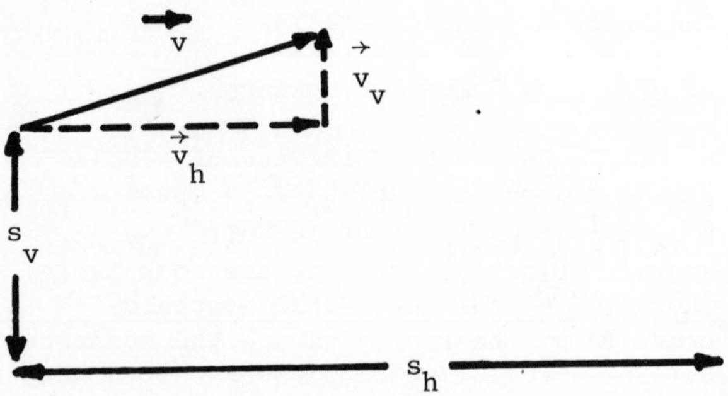

Fig. 2.4

(a) The rock travels a certain distance above the building and returns the same height. At this instant of time it is 24 m above the surface of the earth. Since the distance it moves above the building equals the distance it travels back to 24 m above the earth's surface, the net distance moved during this time is zero. The net displacement during the total time the rock is in the air is -24 m. We use the minus sign because we chose upward as positive. For the vertical motion,

$$-24 \text{ m} = (7 \text{ m s}^{-1})t + 1/2\ (-10 \text{ m s}^{-2})\ t^2$$

or

$$5t^2 - 7t - 24 = 0,$$

where t is in s. Using the solution to a quadratic equation, we find

$$t = \frac{7 \pm \sqrt{49 + 480}}{10} = \frac{7 \pm 23}{10}$$

$$t = 3 \text{ s}$$

The solution for t that gives a negative time has no physical significance.

(b) $s_h = (v_o)_h t = (24 \text{ m s}^{-1})(3 \text{ s}) = 72 \text{ m}$

(c) $v_h = (v_o)_h = 24 \text{ m s}^{-1}$

$v_v = (v_o)_v + a_v t$

$= 7 \text{ m s}^{-1} + (-10 \text{ m s}^{-2})(3 \text{ s})$

$= -23 \text{ m s}^{-1}$

An alternative method for finding the time is to first find the time for the rock to reach its highest point. At this point $v_h = 0$. If we let the time be t_1, then

$v_h = (v_o)_h + at_1$

$0 = 7 \text{ m s}^{-1} + (-10 \text{ m s}^{-2})t_1$

or

$t_1 = 0.7 \text{ s}$

In time t_1, the rock travels a vertical distance

$(s_1)_v = (v_o)_v t_1 + 1/2\ a_v t_1^{\ 2}$

$= (7 \text{ m s}^{-1})(0.7 \text{ s}) + 1/2(-10 \text{ m s}^{-2}) \times$
(0.49 s^2)

$= 2.45 \text{ m}$

It is now a distance $(s_2)_v = 26.45$ m above the ground with an instantaneous velocity of zero. Since $(s_2)_v = 1/2\ a_v t_2^2$, the additional time t_2 for the rock to reach the ground is

$$t_2 = \sqrt{2(s_2)_v / a_v}$$

$$= \sqrt{(2 \times 26.45\text{m})/(10 \text{ m s}^{-2})}$$

$$= 2.3 \text{ s}$$

In the above we took the positive direction to be downward. The total time the rock is in the air is

$$t = t_1 + t_2$$

$$= 0.7 \text{ s} + 2.3 \text{ s} = 3.0 \text{ s}$$

Problems

1. When an object moves with a constant velocity, it

   (a) moves equal distances in equal time intervals

   (b) moves increasing distances in equal time intervals as time increases

2. An object moves with a constant acceleration of 10 m s^{-2}. If its initial velocity is zero, its average velocity during 4 s is

   (a) 10 m s^{-1}

   (b) 20 m s^{-1}

   (c) 40 m s^{-1}

3. In Problem 2, the instantaneous velocity at t = 4 s is

   (a) 10 m s^{-1}

   (b) 20 m s^{-1}

   (c) 16 m s^{-1}

4. An object with an initial velocity of zero experiences an acceleration of 4 m s^{-2}. When it has a velocity of 4 m s^{-1}, it has traveled a distance of

   (a) 2 m

   (b) 4 m

   (c) 8 m

   (d) 16 m

Fig. 2.5

5. Figure 2.5 is a time-position record of an object at equal time intervals. The object

   (a) moves with constant velocity

   (b) experiences a positive acceleration

   (c) experiences a negative acceleration

6. A person walks 24 miles east and then 7 miles north. The person's resultant displacement is

   (a) 31 miles, 16° north of east

   (b) 25 miles, 16° north of east

   (c) 25 miles, 16° south of east

7. An object has an initial velocity of 20 m s^{-1} at t = 0. The acceleration of the object from t = 0 to t = 6 s is shown in Fig. 2.6(a). In Fig. 2.6(b) sketch the velocity of the object as a function of time. Label the values on the velocity axis.

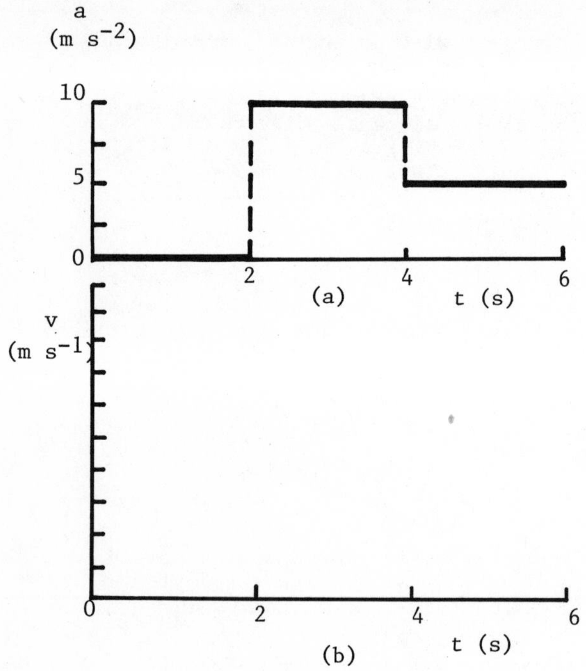

Fig. 2.6

8. In Problem 7, the distance moved by the object in 6 s is

   (a) 70 m

   (b) 150 m

   (c) 190 m

   (d) 210 m

9. A ball is thrown with a speed of 10 m s^{-1} at an angle of 53° with the horizontal. The maximum height reached by the ball is

   (a) 1.8 m

   (b) 3.2 m

   (c) 5.0 m

10. In Problem 9, the horizontal distance moved by the ball is

   (a) 3.2 m

   (b) 4.8 m

   (c) 8.0 m

CHAPTER 3

Dependence of Acceleration on Force and Mass

Objectives

You should

1. identify force as a property of the environment and mass as a property of the object

2. learn how to isolate an object and determine the forces acting on the object

3. realize that Newton's third law of motion forces act on different objects

4. recognize that a horizontal force produces a horizontal acceleration and a vertical force produces a vertical acceleration

Trouble Spots

When you double the force you apply to an object, you may or may not double its acceleration. For example, if a frictional force also acts on an object, the acceleration is not doubled. The product of mass and acceleration equals the net force, not an individual force on the object.

Inertial mass is defined such that the ratio of the inertial mass of two objects equals the inverse ratio of their accelerations, when they experience the same net force. A 2 kg-object experiences one-half the acceleration of a 1-kg object, when they are subjected to the same force.

When a horse pulls a cart, the force F_{hc} exerted by the horse on the cart equals the force F_{ch} of the cart on the horse.

Only one of these forces, f_{ch}, helps determine the motion of the horse, because F_{hc} acts on the cart not on the horse.

Illustrative Examples

1. In Fig. 3.1 we plot the velocity of two objects as a function of time. Comment on the magnitude of the net force acting on the objects and the direction of the force.

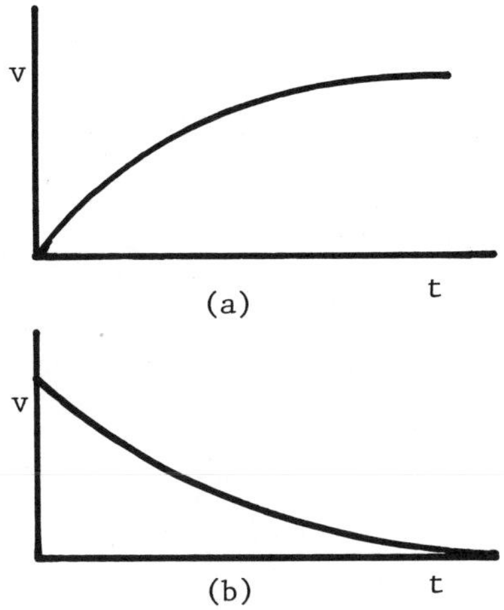

Fig. 3.1

For Fig. 3.1(a), the slope of v vs t (the acceleration) decreases with time. Thus the net force acting on the object decreases with time. The magnitude of the velocity, however, increases with time so the net force must always be in the direction of the motion.

In Fig. 3.1(b), the slope of the velocity versus time decreases with time so the net force must decrease with increasing time. Since the object is slowing down, the net force must be opposite to the direction of the motion.

2. A person exerts a force $F_{pr} = 3.6$ N on a rope of mass $m_r = 0.2$ kg. The rope is attached to a box of mass $m_b = 1.8$ kg,

that rests on a horizontal frictionless table as shown in Fig. 3.2(a). Find (a) the acceleration of the box (b) the force F_{br} of the box on the rope and (c) the force F_{rb} of the rope on the box.

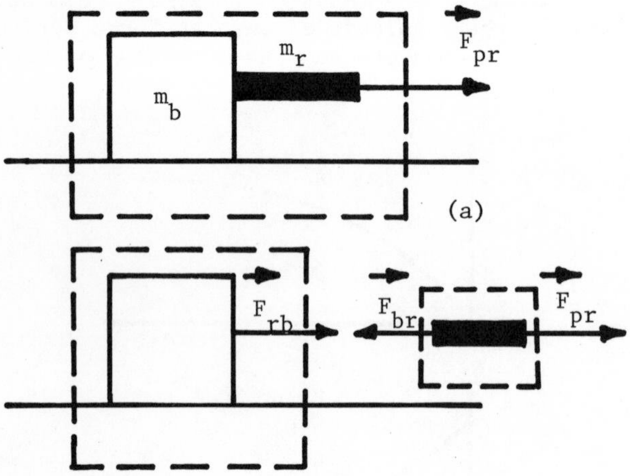

Fig. 3.2

(a) First isolate the total system of mass $m = m_r + m_b = 2.0$ kg, as shown in Fig. 3.2(a) and set the net external force acting on the total system = ma. Notice that F_{rb} and F_{br} are <u>internal</u> forces to this system.

$$\text{Net force} = F_{pr} = ma$$
$$3.6 \text{ N} = (2 \text{ kg})\, a$$
$$1.8 \text{ m s}^{-2} = a$$

(b) To find F_{br} isolate the rope as shown in Fig. 3.2(b). The external forces for this system are F_{pr} and F_{br}.

$$\text{Net force} = F_{pr} - F_{br} = m_r\, a$$
$$3.6\text{N} - F_{br} = (0.2 \text{ kg})(1.8 \text{ m s}^{-2})$$
$$F_{br} = 3.24 \text{ N}$$

(c) To find F_{rb} isolate the box as shown in Fig. 3.2(b). The external force for this system of mass m_b is F_{rb}.

$$\text{Net force} = F_{rb} = m_b a$$
$$= (1.8 \text{ kg})(1.8 \text{ m s}^{-2})$$
$$= 3.24 \text{ N}$$

Notice that F_{br} is equal in magnitude and opposite in direction to F_{rb}.

3. In Problem 2, assume that the box is pulled by a string of negligible mass. Find (a) the acceleration of the box, (b) the force F_{bs} of the string on the box,

and (c) the force F_{sb} of the string on the box.

(a) Now

$$\text{Net force} = F_{ps} = (m_s + m_b)a$$
$$3.6 \text{ N} = (0 + 1.8 \text{ kg})a$$
$$2 \text{ m s}^{-2} = a$$

(b) For the string,

$$\text{Net force} = F_{ps} - F_{bs} = m_s a$$
$$3.6 \text{ N} - F_{bs} = (0)(2 \text{ m s}^{-2})$$
$$F_{bs} = 3.6 \text{ N}$$

(c) For the box,

$$\text{Net force} = F_{sb} = m_b a$$
$$= (1.8 \text{ kg})(2 \text{ m s}^{-2})$$
$$= 3.6 \text{ N}$$

For a massless string, you may consider that the force exerted by the person is transmitted through the string to the object without dimunition.

4. A girl pulls on a massless rope with a force of 20 N at an angle of 37° with the horizontal, as shown in Fig. 3.3. The rope is attached to a box of mass

$m = 4$ kg. The box moves over a horizontal surface that exerts a retarding frictional force $F_f = 4$ N. Find the acceleration of the box.

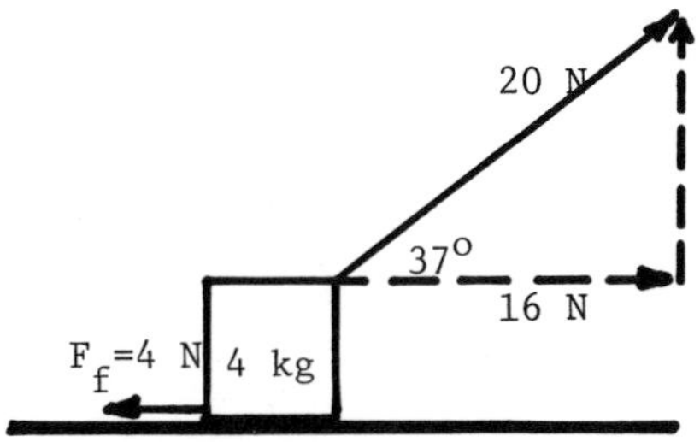

Fig. 3.3

Since the box moves in a horizontal direction, you must find the net force acting in a horizontal direction. The horizontal component of the applied force is

$$F_h = 20 \text{ N} \cos 37^\circ$$

$$= 20 \text{ N} \times 4/5$$

$$= 16 \text{ N}$$

The net force horizontally is

$$F_h - F_f = ma$$

$$16 \text{ N} - 4 \text{ N} = (4 \text{ kg})a$$

$$3 \text{ m s}^{-2} = a$$

Problems

1. A constant net force of 4 N acts on an object of mass $m = 2$ kg. The acceleration of the object is

   (a) zero

   (b) 0.5 m s^{-2}

   (c) 2 m s^{-2}

2. A constant net force gives an object an acceleration a. When the net force is doubled

   (a) the acceleration is halved

   (b) the acceleration remains the same

   (c) the acceleration increases, but is not doubled

   (d) the acceleration is doubled

3. Object A has a mass of 4 kg. Object B has a mass of 2 kg. A net force F is first applied to object A and then to object B. The positions of the objects are measured at equal time intervals. The average velocities are calculated and plotted as a function of time. The slope of the average velocity versus time for object A is

   (a) less than that for object B

   (b) equal to that for object B

   (c) greater than that for object B

4. A man exerts a constant horizontal force of 30 N to give an object of mass m = 5 kg an acceleration of 5 m s^{-2} along a level surface. The frictional force acting on the object is

   (a) zero

   (b) 5 N

   (c) 15 N

5. A man finds that he must exert a horizontal force of 10 N to move an object of mass $m = 4$ kg along a level surface with a constant velocity. To give the object an acceleration of 4 m s^{-2}, he must exert a force of

   (a) 10 N

   (b) 16 N

   (c) 26 N

6. An initially at rest object of mass m = 2 kg is pushed along a horizontal frictionless surface with a constant horizontal force of 4 N. When the object moves a distance of 16 m, the velocity of the object is

(a) zero

(b) 4 m s^{-1}

(c) 8 m s^{-1}

(d) 16 m s^{-1}

(e) 64 m s^{-1}

7. An object of mass m = 2.5 kg experiences a force of 12 N toward the east and 5 N toward the north. No other forces act on the object. The acceleration of the object is

(a) 5.2 m s^{-2}, 22.5^o north of east

(b) 5.2 m s^{-2}, 22.5^o south of east

(c) 5.2 m s^{-2}, 67.5^o north of east

(d) 6.8 m s^{-2}, 22.5^o north of east

8. A ball of mass m = 1.0 kg rolls across a frictionless level table with a speed of 5 m s^{-1} at an angle of 37° north of east. At t = 0, an extensive blower system is turned on that exerts a constant force of 4 N toward the north. The ball travels 6 m east in time t equal to

(a) $\sqrt{3}$ s

(b) 1.2 s

(c) 1.5 s

(d) 2.0 s

9. At time t in Problem 8, the velocity of the ball toward the north is

(a) 6.0 m s^{-1}

(b) 9.0 m s^{-1}

(c) 11 m s^{-1}

(d) 12 m s^{-1}

10. At time t of Problem 8, the total distance moved by the ball is

(a) 7.5 m

(b) 10.8 m

(c) 15.0 m

(d) 16.5 m

Sample Quiz for Chapters 1 through 3

In all cases take $g = 10\ m\ s^{-2}$

1. A boy exerts a force of 25 N to row a boat of mass m = 20 kg at a constant speed of 1.0 m s^{-1} in a straight line. The frictional force exerted by the water on the boat is

   (a) Zero

   (b) 5 N

   (c) 25 N

2. A force of 3 N acts on an object to the east and a force of 4 N acts on the object to the north. In order for the object to move with a constant velocity, there must also be a force F of

   (a) Zero

   (b) 5 N, 53° north of east

   (c) 5 N, 53° south of west

   (d) 5 N, 37° south of north

   (e) 7 N, 53° north of east

3. Figure 1a is a plot of the velocity of a particle as a function of time. In Fig. 1b, plot the acceleration as a function of time. Label the values on the acceleration axis.

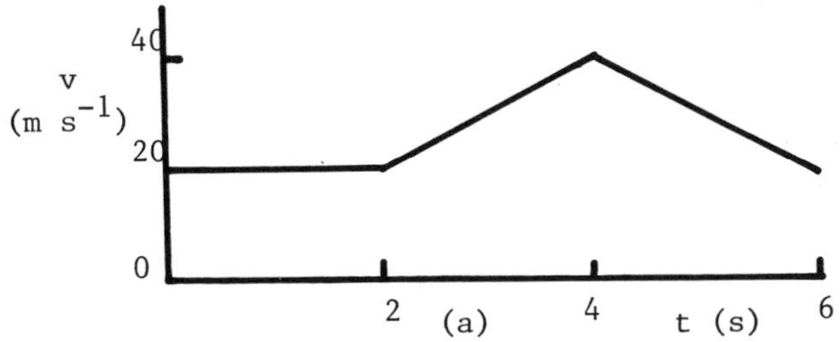

(a)

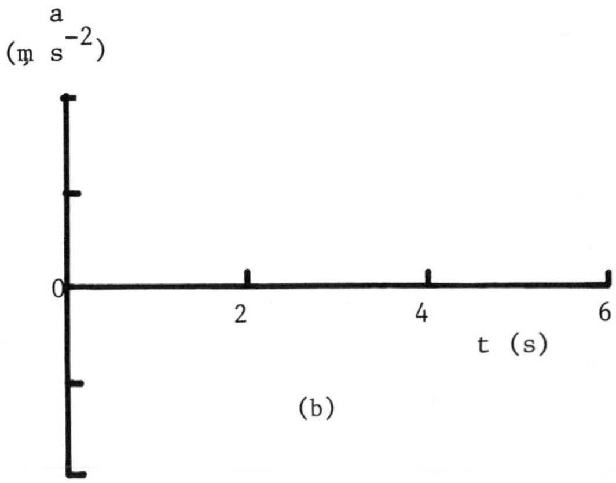

(b)

Fig. 1

4. The distance moved by the object in Problem 4 in 6 s is

   (a) 40 m

   (b) 80 m

   (c) 100 m

   (d) 120 m

   (e) 160 m

   (f) 180 m

5. A person walks at a speed of 13 mph, 22.5^o north of west for 1 hr and then walks at a speed of 12.5 mph, at an angle of 22.5^o north of east for 2 hours. The average velocity for the trip is

(a) 12.7 m s^{-1}, 6.5° north of west

(b) $4\sqrt{2}$ m s^{-1}, 45° north of east

(c) $4\sqrt{10}$ m s^{-1}, 18.4° north of east

6. A bullet is fired with a velocity v_o = 17 m s^{-1} at an angle of 28° with the horizontal from the top of a building that is 85 m high. The horizontal distance that the bullet lands from the building is

(a) 14.45 m

(b) 40 m

(c) 75 m

(d) 150 m

7. The magnitude of the vertical velocity of the bullet in Problem 6 when it hits the ground is

(a) 10 m s^{-1}

(b) $10\sqrt{17}$ m s^{-1}

(c) 42 m s^{-1}

(d) 50 m s^{-1}

8. A person pulls on a rope in a horizontal direction with a force of 12 N and gives a box of mass m = 3 kg an acceleration of 3 m s^{-2} along a horizontal frictionless surface. The mass of the rope is

(a) Zero

(b) 1 kg

(c) 3 kg

9. In Problem 8 the force of the box on the rope is

(a) Zero

(b) 9 N

(c) 12 N

(d) 21 N

10. Three blocks are connected by essentially massless strings as shown in Fig. 2. The masses of the blocks are m_1 = 3 kg, m_2 = 6 kg, and m_3 = 9 kg. The horizontal force applied T_3 = 36 N. The frictional force acting on m_3 is 4.5 N, on m_2 is 3 N, and on m_1 is 1.5 N. The tension T_2 in the rope is

(a) 4.5 N

(b) 9.0 N

(c) 18 N

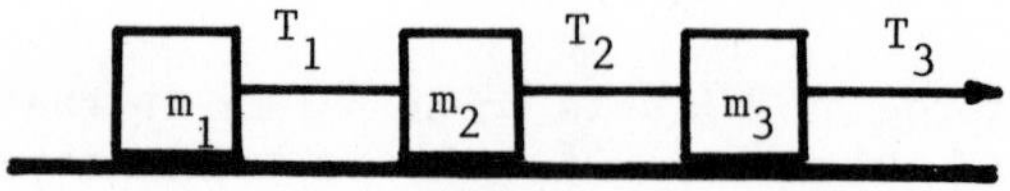

Fig. 2

CHAPTER 4. A Unifying Theory--Newton's Laws

Objectives

You should learn

1. to vary one variable at a time as you test for the dependence of a quantity on a number of variables

2. to recognize mg as one of the forces that acts on an object

3. to draw force diagrams isolating all the forces acting on an object

4. to separate the components of the weight parallel and perpendicular to an incline plane in analyzing motion up or down a plane

5. to identify the centripetal acceleration of an object moving with speed v in a circle of radius r as v^2/r

6. to recognize that an object vibrates with simple harmonic motion when acted upon by a force that is directly proportional, and opposite in direction, to the object's displacement

7. the meaning of the descriptive terms for simple harmonic motion, that is, amplitude A, period T, and frequency f

8. the relation between the displacement s and the acceleration a at any instant for simple harmonic motion, that is, $s/a = -T^2/4\pi^2$

9. to apply the equation of motion for simple harmonic motion,

$$s = A \cos 2\pi t/T$$

Trouble Spots

The acceleration of all freely falling bodies near a certain point on the earth's surface is the same. This should tell you that the gravitational force acting on objects of different mass is not the same. Since a large mass resists a force by a greater amount than a small mass, it must experience a greater force to attain the same acceleration.

The weight of an object is mg not g. Weight is a force and should "go" on the left side of the equation,

$$\text{net force} = ma.$$

The acceleration of an object is g only if the only force that acts on the object is the gravitational force.

For motion up or down an inclined plane, the acceleration is either up or down the plane and, therefore, the net force is either up or down the plane. Unless there is a frictional force (that does depend on the normal force), only the component of the weight parallel to the plane helps determine the acceleration of the object.

Many types of forces can produce a centripetal acceleration. There is not some magical centripetal force called mv^2/r. The product of m and v^2/r is simply the right side of the equation,

$$\text{net force} = ma_c$$

where the centripetal acceleration a_c equals v^2/r.

In simple harmonic motion, the acceleration is directly proportional, and in the opposite direction, to the displacement. The magnitude of the acceleration is greatest for maximum displacement from the equilibrium position. While the acceleration is greatest when the displacement equals the amplitude of the motion, the velocity equals zero. Recall that acceleration equals the rate of change of velocity, not the velocity. The minimum velocity is zero. A negative velocity reflects motion in a negative direction, not a velocity smaller than zero.

When an object has a displacement equal to the amplitude A in simple harmonic motion, the acceleration equals

$$-4\pi^2 A/T^2$$

where T is the period, not the time at which this acceleration occurs. For any other displacement s, the acceleration equals

$$-4\pi^2 s/T^2.$$

For example, when the object is at the equilibrium position (s = 0), a = 0.

The equation of motion for simple harmonic motion, when the object has maximum positive displacement at t = 0, is

$$s = A \cos 2\pi t/T.$$

The argument of the cosine is in radians. For example, when t = T/2,

$$s = A \cos \pi$$

Since $\pi = 180^{\circ}$,

$$s = A \cos 180^{\circ}$$

$$= A\ (-1)$$

$$= -A$$

Illustrative Examples

1. Two blocks of mass m_1 = 8 kg and m_2 = 12 kg are connected by a string. A person raises the blocks by applying a force F, as is shown in Fig. 4.1(a). Find (a) the force F to lift the blocks with a constant velocity v = 5 m s^{-1}, (b) the force F' to lift the blocks with a constant acceleration a = 5 m s^{-2}, and (c) the tension T in the connecting string when a = 5 m s^{-2}. Assume that the mass of the string is negligible. Take g = 10 m s^{-2}.

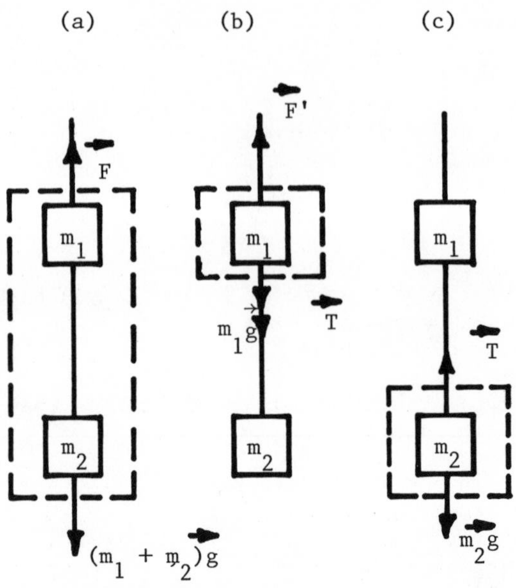

Fig. 4.1

(a) First we isolate the two blocks and look at the forces external to that system. They are F and the weight of the blocks, $(m_1 + m_2)g$ = 200 N. Since the blocks move with a constant velocity, a = 0. Applying Newton's second law, we find

$$\text{Net force} = ma$$

$$F - 200\text{ N} = (20\text{ kg})(0)$$

$$F = 200\text{ N}$$

(b) For an acceleration a = 5 m s^{-2},

$$F' - 200\text{ N} = (20\text{ kg})(5\text{ m s}^{-2})$$

$$F' = 100\text{ N} + 200\text{ N}$$

$$= 300\text{ N}$$

(c) To find the tension T in the string isolate m_1, as shown in Fig. 4.1(b). Now the mass of the isolated system is m_1 = 8 kg and the external forces are F' = 300 N, m_1g = 80 N, and T.

Now applying Newton's second law, we find

$$\text{Net force} = ma$$

$$F' - m_1g - T = m_1a$$

$$300\ \text{N} - 80\ \text{N} - T = (8\ \text{kg})(5\ \text{m s}^{-2})$$

$$300\ \text{N} - 80\ \text{N} - 40\ \text{N} = T$$

$$180\ \text{N} = T$$

Alternatively, you could isolate m_2 = 12 kg as shown in Fig. 4.1(c). Now

$$T - m_2g = m_2a$$

$$T - 120\ \text{N} = (12\ \text{kg})(5\ \text{m s}^{-2})$$

$$T = 60\ \text{N} + 120\ \text{N}$$

$$= 180\ \text{N}$$

2. In Fig. 4.2, m_2 = 4.0 kg, m_1 = 2.0 kg. Find (a) the normal force N, (b) the acceleration a of the system, and (c) the tension T in the string. Assume the string and pulley have essentially zero mass and that the plane is frictionless.

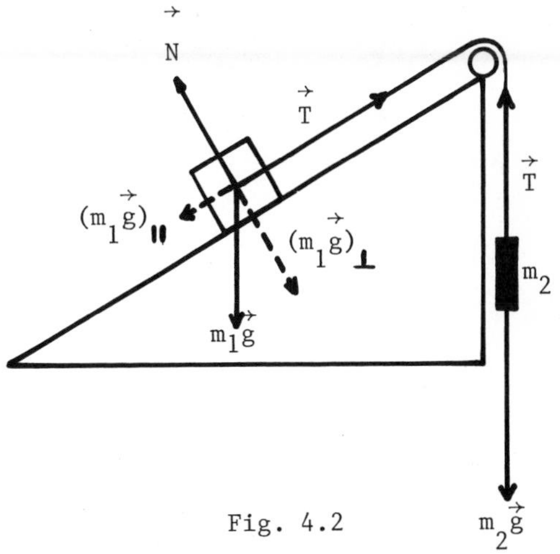

Fig. 4.2

(a) First identify the forces acting on the two objects. The forces acting on m_2 are m_2g = 40 N and the tension T in the string. The forces acting on m_1 are T, m_1g = 20 N, and the normal force N of the plane. The components of the weight m_1g are

$$(m_1g)_{\parallel} = m_1g \sin 30^{\circ}$$

$$= 20\ \text{N} \times 1/2$$

$$= 10\ \text{N}$$

and

$$(m_1g)_{\perp} = m_1g \cos 30^{\circ}$$

$$= 20\ \text{N} \times \sqrt{3}/2$$

$$= 10\sqrt{3}\ \text{N}$$

Since there is no acceleration perpendicular to the plane,

$$(\text{Net force}) = m_1a$$

$$N - (m_1g)_{\perp} = m_1(0)$$

$$N = (m_1 g)_{\perp} = 10\sqrt{3}\ N$$

(b) Isolating the object of mass m_2, we write Newton's second law for this object

$$m_2 g - T = m_2 a$$

or

$$40\ N - T = (4\ kg)a \qquad (1)$$

The object of mass m_1 has the same magnitude of acceleration as the object of mass m_2. This acceleration is up the plane for m_1. So isolating this object, we find

$$(\text{Net force}) = m_1 a$$

$$T - (m_1 g)_{\parallel} = m_1 a$$

$$T - 10\ N = (2\ kg)a \qquad (2)$$

You must solve Equations (1) and (2) simultaneously. If you add the two equations, you find

$$30\ N = (6\ kg)a$$

$$5\ m\ s^{-2} = a$$

(c) Now T can be found from either Eq. (1) or Eq. (2). For example,

$$40\ N - T = (4\ kg)(5\ m\ s^{-2})$$

$$40\ N - 20\ N = T$$

$$20\ N = T$$

Notice that $20\ N - 10\ N = (2\ kg)(5\ m\ s^{-2})$

3. An object of mass m = 0.5 kg is whirled in a vertical circle of radius r = 1 m. The object makes one revolution in $\pi/4$ seconds. Find (a) the frequency of the motion, (b) the speed of the object (c) the centripetal acceleration, (d) the tension T in the string when the object is at its lowest point, (e) the tension T' in the string when the object is at its highest point.

(a) The period of the motion is the time to make one complete revolution or $\pi/4$ s.

The frequency f is the reciprocal of the period or $4/\pi\ s^{-1}$.

(b) In one complete period, the object moves a distance equal to the circumference of the circle, $2\pi r$. Since the speed of the object equals the distance moved divided by the time to move that distance

$$v = 2\pi r/\text{period}$$

$$= 2\pi \times 1\ m/(\pi/4\ s)$$

$$= 8\ m\ s^{-1}$$

(c) The centripetal acceleration

$$a_c = v^2/r$$

$$= (64\ m^2\ s^{-2})/(1\ m)$$

$$= 64\ m\ s^{-2}$$

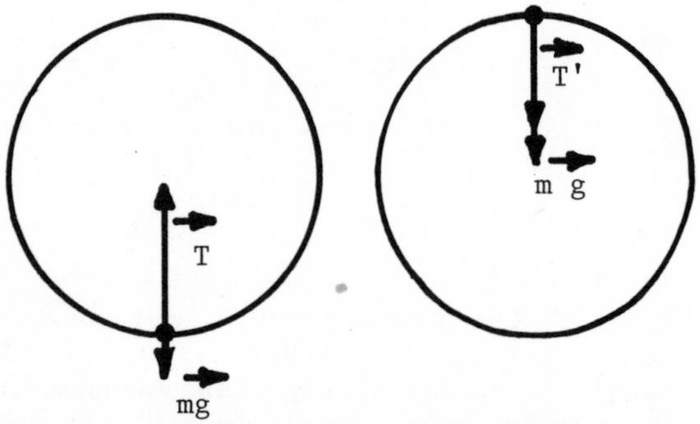

(a) Fig. 4.3 (b)

(d) In Fig. 4.3(a), we look at the forces acting on the object at its lowest point. They are mg = 5 N down and T up. T must be up because, the net force is always in toward the center of the circle and at the bottom of the swing that is up. Applying Newton's second law, we find

$$\text{Net force} = ma$$

$$T - mg = ma$$

$$T - 5\ N = (0.5\ kg)(64\ m\ s^{-2})$$

$$T = 32\ N + 5\ N$$

$$= 37\ N$$

(e) At all points in the motion, the net force acting on the object must equal ma = 32 N. At the highest point, mg supplies a force of 5 N down or in toward the center of the circle. T' must supply an additional force downward. Now

$$T' + 5\ \text{N} = 32\ \text{N}$$

$$T' = 27\ \text{N}$$

4. A mass m = 2.5 kg attached to a spring with spring constant $k = 10\ \text{N m}^{-1}$, vibrates back and forth along a horizontal, frictionless surface with simple harmonic motion. Find (a) the force acting on the mass when it is a distance x = 0.25 m from the equilibrium position, (b) its acceleration when x = 0.25 m, and (c) the period of the motion.

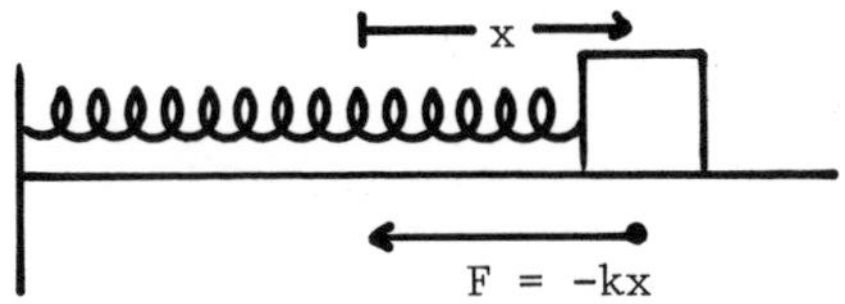

Fig. 4.4

(a) The only force acting on the mass in a horizontal direction is the force due to the spring, as shown in Fig. 4.4.

$$\begin{aligned} F &= -kx \\ &= -(10\ \text{N m}^{-1})(0.25\ \text{m}) \\ &= -\ 2.5\ \text{N} \end{aligned}$$

(b)

$$\begin{aligned} \text{Net force} &= ma \\ -kx &= ma \\ -2.5\ \text{N} &= (2.5\ \text{kg})a \\ -1\ \text{m s}^{-2} &= a \end{aligned}$$

(c)

$$\begin{aligned} \text{Period } T &= 2\pi\sqrt{-x/a} \\ \text{Period } T &= 2\pi\sqrt{(-0.25\ \text{m})/(-1\ \text{m s}^{-2})} \\ &= 2\pi\sqrt{0.25\ \text{s}^2} \\ &= 2\pi\ \ \text{x}\ 0.5\ \text{s} \\ &= \pi\ \text{s} \end{aligned}$$

5. In problem 4, the amplitude of the motion is 0.5 m. (a) Write the equation of motion for the object, if the object has maximum displacement at t = 0. (b) Find the acceleration of the object when t = 0.325 s.

(a) In general,

$$s = A \cos 2\pi t/T$$

for A = 0.5 m and T = π s = period

$$s = (0.5\ \text{m}) \cos 2t$$

(b) For t = 0.325 s,

$$s = (0.5\ \text{m}) \cos 0.65$$

Since 0.65 radians = 37°, at t = 0.325 s

$$\begin{aligned} s &= (0.5\ \text{m}) \cos 37^\circ \\ &= (0.5\ \text{m}) \times 4/5 \\ &= 0.4\ \text{m} \end{aligned}$$

Then

$$\begin{aligned} a &= -4\pi^2\ s/T^2 \\ &= -4\pi^2\ (0.4\ \text{m})/\pi^2\ \text{s}^2 \\ &= -1.6\ \text{m s}^{-2} \end{aligned}$$

Problems

1. The mass of an object is m. Its weight is

(a) m

(b) g

(c) mg

2. To lift an object with a constant velocity v, you must exert a force of

(a) Zero

(b) mg

(c) (mv + mg)

3. A person of mass m stands on a scale on an elevator that moves down with constant velocity v. The reading of the scale is

(a) Zero

(b) m

(c) mg

(d) (mg - mv)

4. A person of mass m stands on a scale on an elevator that accelerates downward with a constant acceleration a. The reading on the scale is

(a) Zero

(b) mg

(c) (mg - ma)

(d) (mg + ma)

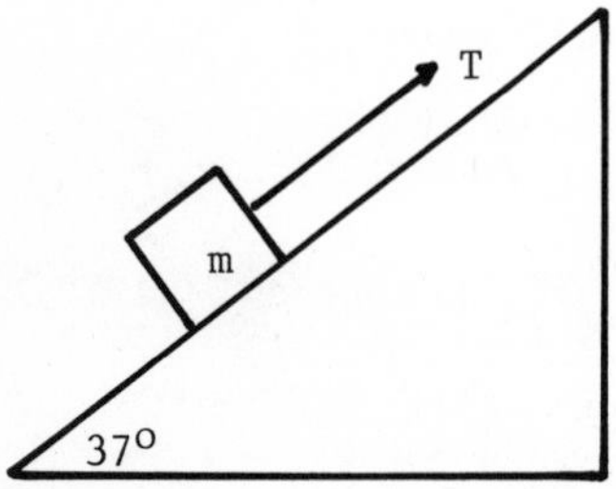

Fig. 4.5

5. A block of mass m = 5 kg is pulled up a frictionless inclined plane at angle of 37° with the horizontal (Fig. 4.5). The acceleration of the block is 3.0 m s^{-2}. The tension in the massless string is

(a) 15 N

(b) 18 N

(c) 45 N

(d) 50 N

6. If the string in Fig. 4.5 is released and the block slides down the plane, its acceleration is

(a) 0

(b) 3/5 g

(c) 4/5 g

(d) g

7. A satellite of mass m revolves around a planet of mass m in a circle of radius r. The velocity of the satellite is

(a) $\sqrt{gr}$

(b) $\sqrt{GM}$

(c) $\sqrt{GM/r}$

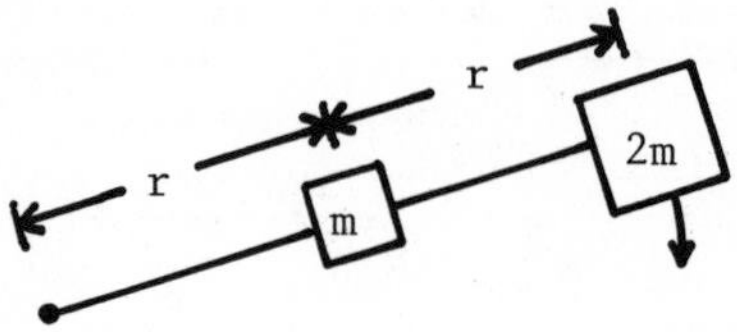

Fig. 4.6

8. Two blocks of mass m and 2 m (Fig. 4.6) on a horizontal frictionless table move with uniform circular motion with period T. The force F_1 exerted by the innermost string is

(a) $4\pi^2 mr/T^2$

(b) $16\pi^2 mr/T^2$

(c) $20\pi^2 mr/T^2$

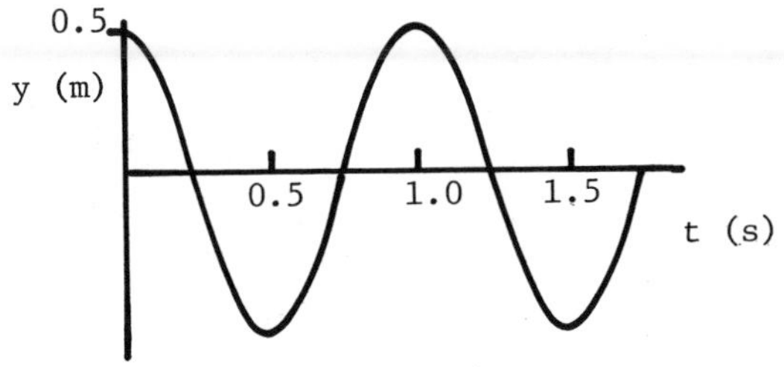

Fig. 4.7

9. Figure 4.7 is a plot of the displacement of an object vibrating with simple harmonic motion as a function of time. The period of the motion is

   (a) 0.25 s

   (b) 0.5 s

   (c) 1.0 s

10. The acceleration of the object in Problem 9 at t = 0.5 s is

    (a) 0

    (b) $-2\pi^2$ m s^{-2}

    (c) $2\pi^2$ m s^{-2}

    (d) $8\pi^2$ m s^{-2}

CHAPTER 5. Forces

Objectives

You should learn

1. the difference between inertial and gravitational mass

2. the direction of forces between like and unlike charges

3. the similarites and differences between gravitational and electric forces

4. the use of the principle of superposition for finding the net force on an object due to two or more other objects

5. the meaning of a north-seeking pole and a south-seeking pole

6. that a current carrying wire carries no net charge

7. the difference between electric and magnetic interactions

8. that the electric interaction takes place between objects with charge whether they are at rest or in motion

9. that the magnetic interaction takes place between charges only when they are in motion

10. that magnetic interactions take place between two magnets, between a magnet and a current carrying wire, and between two current carrying wires

11. how to apply the right hand rule to find the direction a north-seeking pole is

urged due to a long current carrying wire

12. the direction of forces between two current carrying wires

Trouble Spots

The inertial mass of an object is the property of an object to resist an acceleration. The gravitational mass of an object is its ability to take part in a gravitational interaction. Because of the equivalence of inertial and gravitational mass, all freely falling objects fall with the same constant acceleration. In addition, because of this equivalence, it is impossible to tell whether you are experiencing a gravitational interaction or are in an accelerated system.

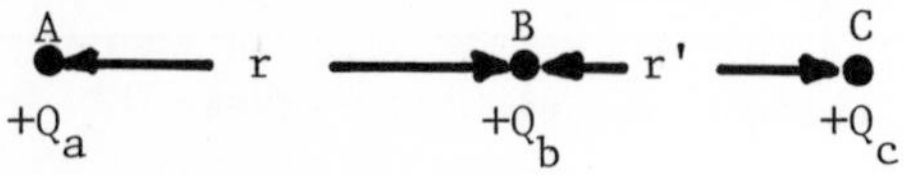

Fig. 5.1

Forces are repulsive between like charges and attractive between like charges. Do not make the mistake, however, of substituting signs into Coulomb's law,

$$F_e = k_e qq'/r^2.$$

For example, in Fig. 5.1, Q_a repels Q_b and $-Q_c$ attracts Q_b. If you use signs of the charges in Coulomb's law you might say there was a positive force of a on b and a negative force of c on b and subtract the magnitudes of the two forces to find the net force on b. In reality, both a and c exert a force on b to the right, and the net force on b is the sum of the forces due to a and c.

The end of a compass that points to geographic north is called a north-seeking pole. Since unlike poles attract that means the earth acts as though it had a south-seeking pole in the northern hemisphere.

While current carrying wires have a net flow of charge, they do not have a net charge. For this reason there is no electric force between two current carrying wires. (This statement must be modified for observers moving relative to the wire.)

When applying the right hand rule to find the direction of deflection of a north-seeking pole, your fingers must be at the position of the pole. Of course, your fingers point in different directions at different positions.

No one knows "why" two current carrying wires attract each other anymore than anyone knows why two objects that possess mass attract each other or two charges of the same sign repel each other. The direction of the forces is determined by experiment. Remember the role of the physicist is to describe physical phenomena, not to answer "why" it happens.

Illustrative Examples

1. In Fig. 5.1 let $Q_a = +2 \times 10^{-6}$ C, $Q_b = 10^{-6}$ C, $Q_c = -10^{-6}$ C, $r = 3 \times 10^{-2}$ m and $r' = 10^{-2}$ m. Find the force (direction) and magnitude) on b.

   The force of a on b is

$$F_{ab} = \frac{k_e q_a q_b}{r^2}$$

$$= \frac{9 \times 10^{9}\ \text{N-m}^2\ \text{C}^{-2} \times 2 \times 10^{-6} \times 10^{-6}\ \text{C}}{9 \times 10^{-4}\ \text{m}^2}$$

$$= 20\ \text{N to the right}$$

   The force of c on b is

$$F_{cb} = \frac{k_e q_c q_b}{r'}$$

$$= \frac{9 \times 10^{9}\ \text{N-m}^2\ \text{C}^{-2} \times 10^{-6}\ \text{C} \times 10^{-6}\ \text{C}}{10^{-4}\ \text{m}^2}$$

$$= 90\ \text{N to the right}$$

   The resultant force on b is 110 N to the right.

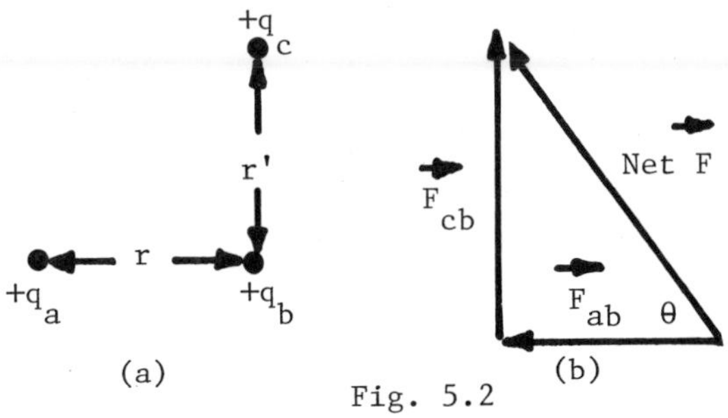

Fig. 5.2

2. In Fig. 5.2(a) let $q_a = -3 \times 10^{-6}$ C, $q_b = 10^{-6}$ C, $q_c = -4 \times 10^{-6}$ C and $r = r' = 10^{-2}$ m. Find the resultant force (direction and magnitude) on b.

$$F_{ab} = \frac{9 \times 10^9 \text{ N-m}^2 \text{ C}^{-2} \times 3 \times 10^{-6} \text{ C} \times 10^{-6}\text{C}}{10^{-4} \text{ m}^2}$$

$= 270$ N, to the left

$$F_{cb} = \frac{9 \times 10^9 \text{ N-m C}^{-2} \times 4 \times 10^{-6}\text{C} \times 10^{-6} \text{ C}}{10^{-4} \text{ m}^2}$$

$= 360$ N, up

In Fig. 5.2(b), we show the vector addition of F_{ab} and F_{cb}.

$$\text{Net F} = \sqrt{(270)^2 + (360)^2} \text{ N}$$

$$= \sqrt{(90 \times 3)^2 + (90 \times 4)^2} \text{ N}$$

$$= 90 \text{ N} \sqrt{(3)^2 + (4)^2}$$

$$= 90 \text{ N} \sqrt{25}$$

$$= 450 \text{ N}$$

$$\tan\theta = 360/270 = 4/3$$

$$\theta = 53^o$$

3. Charge $q = 10^{-6}$ C and $q' = 2 \times 10^{-6}$ C move parallel to each other with a speed of $v = 6 \times 10^6$ m s^{-1}. The initial distance of separation is 10^{-2} m. Find the electric force between q and q'.

The electric force between q and q' is independent of the particles' velocity.

$$F_e = \frac{9 \times 10^9 \text{ N-m}^2 \text{ C}^{-2} \times 10^{-6} \text{ C} \times 2 \times 10^{-6} \text{ C}}{10^{-4} \text{ m}^2}$$

$= 180$ N

The direction of the force is along the line connecting the two charges and it is repulsive.

4. In Problem 3, find (a) the magnetic force between q and q' for $r = 10^{-2}$ m and (b) the ratio of the magnetic and electric forces.

(a) The magnetic force does depend on the velocity of the charges.

$$F_m = \frac{k_m qvq'v'}{r^2} = \frac{k_m qq'v^2}{r^2}$$

$$= \frac{10^{-7}\text{N C}^{-2} \text{ s}^2 \times 2 \times 10^{-12} \text{ C}^2 \times 36 \times 10^{12}\text{m}^2\text{s}^{-2}}{10^{-4} \text{ m}^2}$$

$= 72 \times 10^{-3}$ N

The magnetic force is an attractive force along the line connecting the two charges.

(b) $$\frac{F_m}{F_e} = \frac{72 \times 10^{-3} \text{ N}}{180 \text{ N}}$$

$$= 4 \times 10^{-4}$$

For $v = 6 \times 10^6$ m s^{-1} and the speed of light $c = 3 \times 10^8$ m s^{-1}, notice that

$$(v/c)^2 = (36 \times 10^{12}/9.0\ 10^{16}) = 4 \times 10^{-4}$$

5. The speed of the conduction electrons in a current carrying wire is small compared to the speed of light. If a current carrying wire had a net charge would you be able to detect magnetic forces between them?

As seen in Problem 4, for v << c, the magnetic force is much smaller than the electric force. The net charge of current carrying wires must be zero or we would not be able to detect magnetic forces between them.

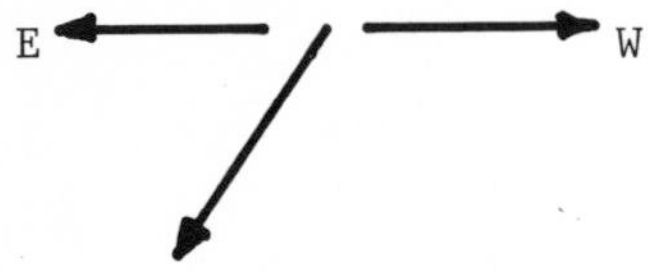

Fig. 5.3

6. Imagine that out of the page pointed toward the north (Fig. 5.3). If a compass needle is placed below a wire with a current toward the north, in what direction will the north-seeking pole be deflected?

Since you move in a clockwise direction from north to east, in Fig. 5.3 east is to the left and west is to the right. If you point the thumb of your right hand out of the page and curl your fingers so that they are below your thumb, you find that your fingers point to the right or to the west.

Problems

1. An object of gravitational mass $m_g = 2$ kg rests on a scale on the earth's surface where $g = 9.8$ m s^{-2}. The reading on the scale is

(a) Zero

(b) 9.8 N

(c) 19.6 N

(d) 29.4 N

2. An object of inertial mass $m_I = 2$ kg is in a spaceship in interplanetary space far from any massive objects. If the object is on the "floor" of the spaceship that accelerates up at $a = 9.8$ m s^{-2}, the reading on the scale is

(a) Zero

(b) 9.8 N

(c) 19.6 N

(d) 29.4 N

3. The equivalence of inertial and gravitational mass tells us that it is impossible to tell

(a) whether you are at rest or moving with a constant velocity

(b) whether you are accelerating or participating in a gravitational interaction

4. Imagine that there is a planet where the gravitational mass of an object is one-fourth its inertial mass. A person on earth suspends an object of inertial mass $m_I = 0.1$ kg from a string and measures the period of the pendulum. A person on the imaginary planet also suspends an object of inertial mass $m_I = 0.1$ kg from a string of the same length and measures its period T'. Which of the following is true?

(a) T = T'/4

(b) T = T'/2

(c) T = T'

(d) T = 2T

(e) T = 4T'

5. A glass rod is rubbed with a silk cloth and acquires a positive charge. It does this because

(a) positive charge leaves the silk and is transferred to the rod

(b) all of the electrons from the glass rod are transferred to the silk

(c) some of the electrons from the glass rod are transferred to the silk

6. Charge A exerts a force F on charge B when they are separated by a distance r. When the distance between them is 2r, the force F' of A on B is

(a) F/4

(b) F/2

(c) F

(d) 2F

(e) 4F

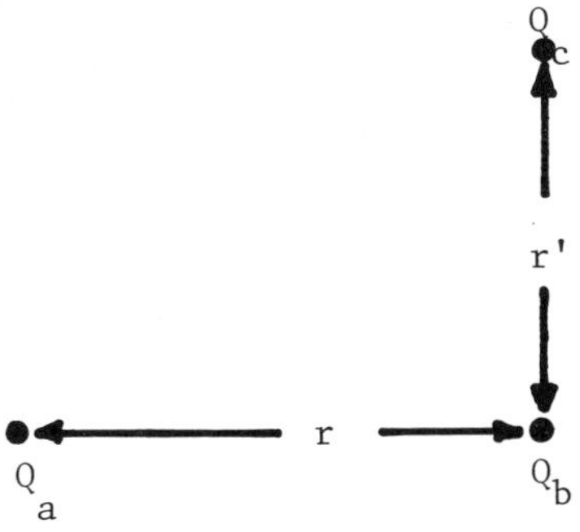

Fig. 5.4

7. In Fig. 5.4, the charge at A, $Q_a = -125 \times 10^{-6}$ C, the charge at B, $Q_b = 1/9 \times 10^{-7}$ C, and the charge at C, $Q_c = -144 \times 10^{-6}$ C. The distance r = 5×10^{-2} m and the distance r' = $2\sqrt{3} \times 10^{-2}$ m. The resultant force on the charge at B is

(a) 7 N, toward C

(b) 13 N, 67.5^o above line from B to A

(c) 13 N, 67.5^o above line from A to B

(d) 17 N toward C

8. A stationary point charge experiences a net force when it is near

(a) another charge

(b) a magnet

(c) a current carrying wire

9. When charge q moving with speed v moves parallel to charge q' moving with speed v and they are separated by distance r, it experiences a magnetic force F. If the speeds are doubled and the distance of separation is halved, the magnetic force F' on q is

(a) F/2

(b) F

(c) 4F

(d) 8F

(e) 16F

10 A current carrying wire is in a north-south direction. The north-seeking pole placed above the wire is deflected toward the west. The current in the wire is

(a) toward the north

(b) toward the south

CHAPTER 6. Fields

Objectives

You should learn

1. to differentiate between the action at a distance approach and the field approach

2. the definition of an electric field at a point as the force on a test charge q' placed at that point divided by the magnitude of the test charge

3. the direction of an electric field as that in which a positive test charge is urged

4. the use of the density of field lines as a measure of the magnitude of the field

5. that the direction of the force on a positively charged particle in an electric field is in the direction of the field and the force on a negatively charged particle is opposite to the direction of the electric field

6. to use the principle of superposition to find fields due to a distribution of charge or current elements

7. the direction of magnetic field lines due to various distributions of current elements

8. to use the first and second right hand "rules"

9. that the force on a charged particle moving in a magnetic field depends on its charge, its velocity v, the strength of the magnetic field B, and the sine of the angle between $\vec{v}$ and $\vec{B}$

10. to distinguish between Problem I and Problem II, as described in the text, and how to apply the approach

11. for magnetic forces to distinguish between up and down and in and out of the page

12. to find the motion of particles in electric and/or magnetic fields

13. that since $C\ s^{-1} = A$, $N(A\text{-}m)^{-1} = N\text{-}s(C\text{-}m)^{-1}$

Trouble Spots

When you are asked to find the electric field due to two point charges, do not find the force of one charge on the other. An electric field is force per unit charge and has the unit of $N\ C^{-1}$, not N.

The electric field for a point charge Q at a distance r is

$$E = k_e Q/r^2.$$

This is not a correct expression for the electric field due to a distribution of charges. For example, the electric field due to an infinite plane of charge is independent of the distance from the plane.

When you are finding the electric field due to a number of charges, use the principle of superposition. Do not determine the direction of the field at a point for a +Q by using a plus sign for the field and for a -Q by using a negative sign for the field. Rather decide if the field due to the +Q and the -Q is to the right or to the left, up or down, in or out of the page.

The magnetic field lines due to a long current carrying wire make concentric circles with the wire. The magnetic field line pattern for a long current carrying wire is not the same as a current carrying loop. A long current carrying wire does not act like a north-seeking pole nor a south-seeking pole.

While the force on a charged particle in an electric field is in the same or opposite direction to the electric field, the magnetic force on a charged particle, moving with velocity v, is always perpendicular to the plane that contains the magnetic field $\vec{B}$ and $\vec{v}$. There is no perversiveness of the physicists in all of this; it is a result of the definition of the magnetic field and experiment. It is even impossible to define the direction of the magnetic field as that in which a moving charge experiences no force because this will happen if the particle moves parallel or anti-parallel to the field.

A charged particle experiences a force in an electric field whether it is at rest or in motion. A charged particle experiences a force in a magnetic field only if it is in motion and if its velocity has some component that is perpendicular to the magnetic field.

When a positive charge moves in an uniform electric field, the direction of the force remains constant: in the direction of the electric field. The acceleration of the charge is constant in the direction of the field. When a positive charge moves in an uniform magnetic field, such that the velocity of the particle is perpendicular to the field, the magnetic force continuously changes direction. The magnetic force does not change the magnitude of the velocity, but it does change its direction. The magnetic force is always perpendicular to the velocity and to the magnetic field.

For no deflection in superposed electric and magnetic fields, the magnetic force on a charged particle must be equal and opposite to the electric force on the charged particle. While the magnetic force must be equal in magnitude and opposite in direction to the electric force, the magnetic field is not equal and opposite to the electric field for no deflection.

Illustrative Examples

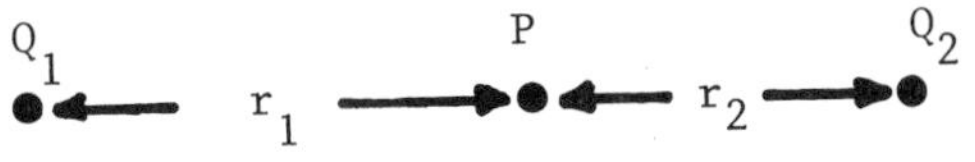

Fig. 6.1

1. In Fig. 6.1, $Q_1 = +16 \times 10^{-6}$ C, $r_1 = 4 \times 10^{-2}$ m, $Q_2 = +9 \times 10^{-6}$ C and $r_2 = 3 \times 10^{-2}$ m. Find the electric field at P.

The electric field due to Q_1 is

$$E_1 = \frac{k_e Q_1}{r_1^2}$$

$$= \frac{9 \times 10^9 \text{ N-m}^2 \text{ C}^{-2} \times 16 \times 10^{-6} \text{ C}}{16 \times 10^{-4} \text{ m}^2}$$

$$= 9 \times 10^7 \text{ N C}^{-1} \text{ to the right}$$

The electric field due to Q_2 is

$$E_2 = \frac{k_e Q_2}{r_2^2}$$

$$= \frac{9 \times 10^9 \text{ N-m}^2 \text{ C}^{-2} \times 9 \times 10^{-6} \text{ C}}{9 \times 10^{-4} \text{ m}^2}$$

$$= 9 \times 10^7 \text{ N C}^{-1} \text{ to the left}$$

The electric field at $P = E = E_1 + E_2 = 0$

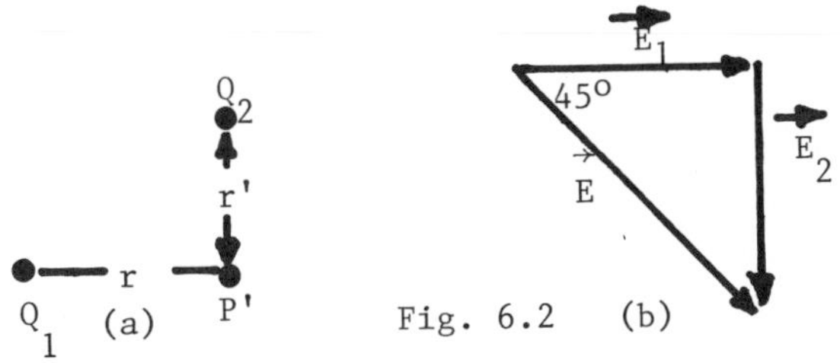

Fig. 6.2

2. The charges of Problem 1 are now rearranged as shown in Fig. 6.2. Find the electric field at P'.

The magnitudes of E_1 and E_2 remain the same because Q_1, Q_2, r_1, r_2 remain the same, but the direction of E_2 changes. E_2 is now down. The resultant field

$$E = \sqrt{(9 \times 10^7)^2 + (9 \times 10^7)^2} \text{ N C}^{-1}$$

$$= 9 \times 10^7 \text{ N C}^{-1} \sqrt{(1)^2 + (1)^2}$$

$$= 9\sqrt{2} \times 10^7 \text{ N C}^{-1}$$

The direction of the electric field is at an angle of 45° below the line from Q_1 to P', as shown in Fig. 6.2(b).

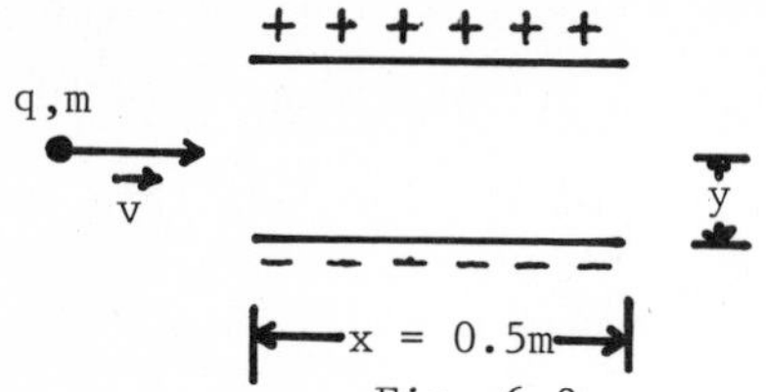

Fig. 6.3

3. A charged particle with $q = +10^{-6}$ C and mass $m = 10^{-10}$ kg traveling with a horizontal speed $v = 10^4$ m s^{-1} enters the uniform electric field $E = 10^4$ N C^{-1} between the parallel plates, shown in Fig. 6.3. Find (a) the electric force on the particle, (b) the acceleration of the particle. The acceleration due to gravity is 9.8 m s^{-2}. (c) Can you ignore the gravitational acceleration? Find (d) the time for the particle to travel through the plates of length x = 0.5 m, (e) the distance y moved downward by the particle at the right edge of the plates.

(a) $F_e = qE$

$= 10^{-6}\ \text{C} \times 10^4\ \text{N C}^{-1}$

$= 10^{-2}$ N

(b) $a = \frac{F_e}{m}$

$= \frac{10^{-2}\ \text{N}}{10^{-10}\ \text{kg}}$

$= 10^8\ \text{m s}^{-2}$

(c) Since 10^8 m s^{-2} is much greater than 9.8 m s^{-2}, we ignore the acceleration due to gravity.

(d) The electric force does not produce an acceleration in the x-direction. The time to travel a distance x = 0.5 m is

$t = x/v$

$= 0.5\ \text{m}/(10^4\ \text{m s}^{-1}) = 0.5 \times 10^{-4}$ s

(e) Since there is no initial velocity in the y-direction,

$y = 1/2\ a_y t^2$

$= 1/2(10^8\ \text{m s}^{-2})(0.25 \times 10^{-8}\ \text{s}^2)$

$= 0.125$ m

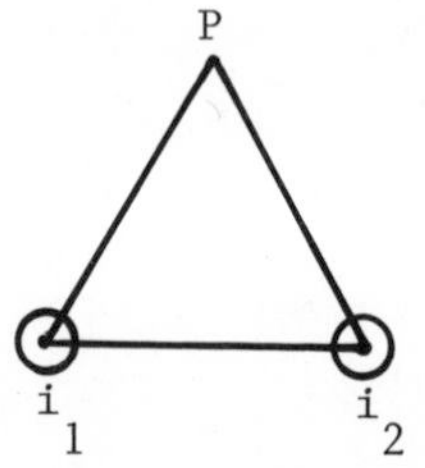

Fig. 6.4(a)

4. Two wires are at the corners at an equilateral triangle, as shown in Fig. 6.4(a). Both wires carry an equal current $i_1 = i_2$ out of the page. The magnitude of the field due to each wire at P, the third corner of the triangle, is 2 N(A-m)$^{-1}$. Find the resultant field at P.

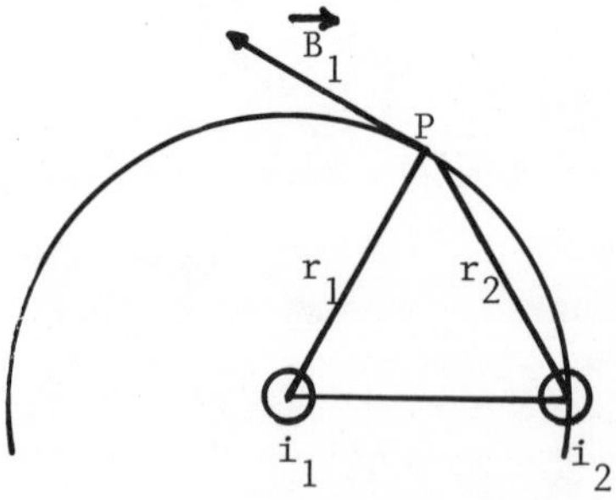

Fig. 6.4(b)

To find the direction of the field due to i_1 draw a circle concentric with i_1 that passes through P, as shown in Fig. 6.4(b). The direction of the field B_1 at P due to i_1 is tangent to the circle or perpendicular to r_1.

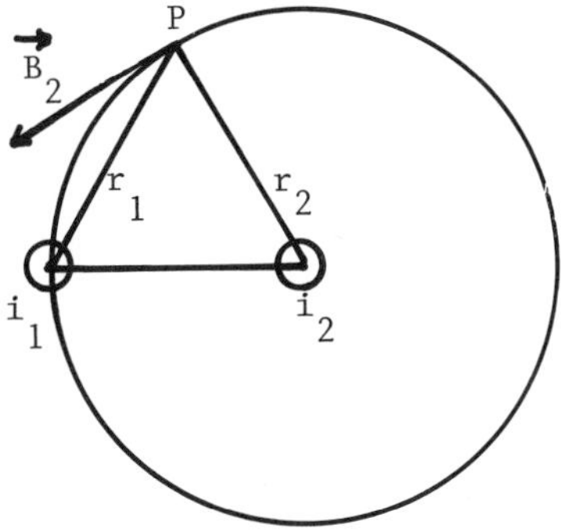

Fig. 6.4(c)

The direction of the field B_2 due to i_2 at P is tangent to the circle at P or perpendicular to r_2, as shown in Fig. 6.4(c).

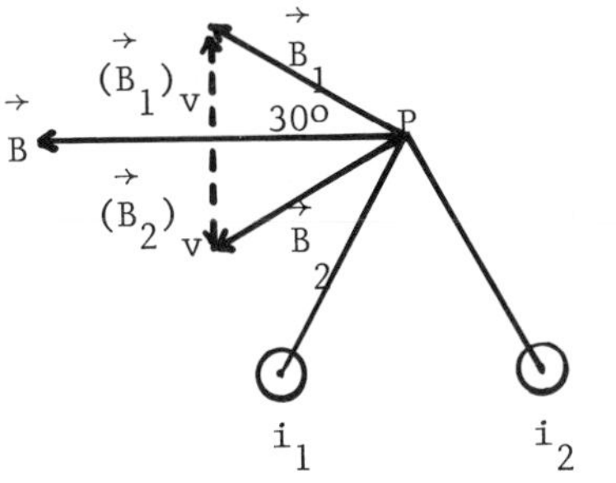

Fig. 6.4(d)

Since $\angle BPi_1$ in Fig. 6.4(d) is an alternate angle to $\angle Pi_1i_2$, $\angle BPi_1$ equals 60° and $\angle B_1PB = \angle BPB_2 = 30^o$. Because $B_1 = B_2$, the vertical components cancel. The resultant field

$$B = 2(B_1)_h = 2(B_2)_h = 2B_1 \cos 30^o$$

$$= 2 \times 2 \text{ N(A-m)}^{-1} \times \sqrt{3}/2$$

$$= 2\sqrt{3} \text{ N(A-m)}^{-1}$$

The direction of B, as shown in Fig. 6.4(d) is to the left.

5. In Problem 4, imagine that a positive charge with $q = 10^{-6}$ C moves out of the page with a velocity of 10^6 m s^{-1}. What is the force (magnitude and direction) on the charge.

The angle between $\vec{v}$ and $\vec{B}$ is 90°.

The magnitude of the force is then

$$F_m = Bqv \sin 90^o$$

$$= 2\sqrt{3} \text{ N(A-m)}^{-1} \times 10^{-6} \text{ C} \times 10^6 \text{ m s}^{-1} \times 1$$

$$= 2\sqrt{3} \text{ N}$$

Using the second right hand rule, you point the thumb of your right hand out of the page, point your fingers to the left, and you find that your palm is ready to push in the direction of the magnetic force, straight down.

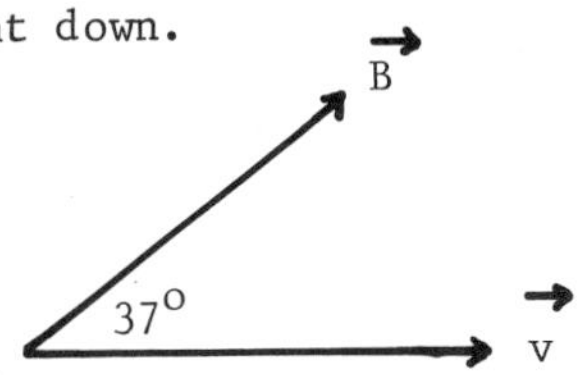

Fig. 6.5

6. A positively charged particle with $q = 10^{-6}$ C moves with a velocity of 10^6 m s^{-1} at an angle of 37° with a magnetic field $B = 10^{-1}$ N(A-m)$^{-1}$, as shown in Fig. 6.5. What force (magnitude and direction) acts on the particle?

$$F_m = qvB \sin \theta$$

$$= 10^{-6} \text{ C} \times 10^6 \text{ m s}^{-1} \times 10^{-1} \text{ N(A-m)}^{-1} \times 3/5$$

$$= 0.06 \text{ N, out of the page}$$

7. A charged particle $q = 10^{-5}$ C and mass $m = 10^{-10}$ kg moves with a velocity $v = 10^6$ m s^{-1} perpendicular to a magnetic field $B = 1$ N(A-m)$^{-1}$. What is the radius of the orbit?

The magnetic field produces a magnetic force

$$F_m = Bqv \sin 90^o$$

$= 1\ N(A\text{-}m)^{-1} \times 10^{-5}\ C \times 10^{6}\ m\ s^{-1} \times 1$

$= 10\ N$

Since this force is always perpendicular to v, it produces a centripetal acceleration so that

$10\ N = mv^2/r$

$r = mv^2/10\ N$

$= (10^{-10}\ kg \times 10^{12}\ m^2\ s^{-2})/10\ N$

$= 10\ m$

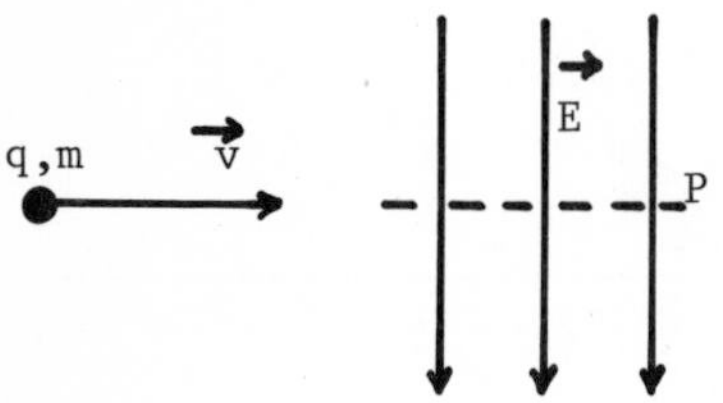

Fig. 6.6

8. A particle of charge +q move with velocity $v = 10^5\ m\ s^{-1}$ into a region of an uniform electric field $E = 10^5\ N\ C^{-1}$. What is the (direction and magnitude of) the magnetic field, perpendicular to v, that will allow the particle to arrive at P (Fig. 6.6) with no deflection?

The electric force due to the electric field

$F_e = qE$

and it is down. For no deflection, the magnetic force F_m must be up. The magnitude of the force

$F_m = qvB$

Thus for no deflection,

$qvB - Eq = 0$

or

$B = E/v$

$= (10^5\ N\ C^{-1})/(10^5\ m\ s^{-1})$

$= 1\ N(A\text{-}m)^{-1}$

For the magnetic force to be up, with a velocity to the right, you find by pointing your thumb in the direction of the velocity (to the right) that your palm will be ready to push up (the direction of the force) if your fingers are into the page (direction of the magnetic field).

9. A particle with charge $q = 10^{-6}$ C moves toward the right with velocity $v = 10^6\ m\ s^{-1}$ in a magnetic field $B = 1\ N(A\text{-}m)^{-1}$. It experiences a force out of the page equal to 0.6 N. What is the direction of the magnetic field?

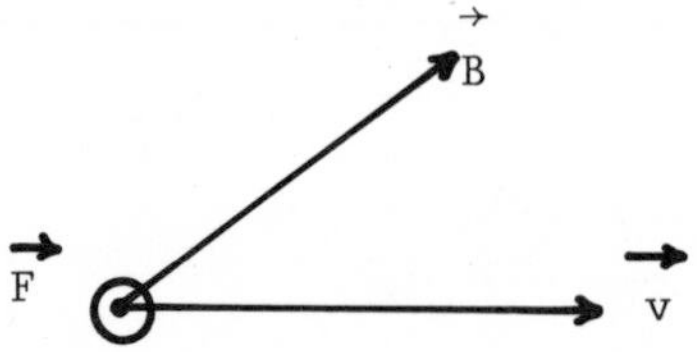

Fig. 6.7

The situation is shown in Fig. 6.7. Since F_m is out of the page and v is in the plane of the page, $\vec{B}$ must lie in the plane of the page because $\vec{F}_m$ is always perpendicular to the plane that contains $\vec{B}$ and $\vec{v}$.

In general,

$F_m = Bqv \sin \theta$

For this case,

$0.6\ N = 1\ N(A\text{-}m)^{-1} \times 10^{-6}\ C \times 10^{6}\ m\ s^{-1} \sin \theta$

$0.6 = \sin \theta$

And $\theta = 37^{o}$

10. A positive charge enters a magnetic field as shown in Fig. 6.8, and the path of the particle is as shown. What is the direction of the magnetic field. Assume

that it is perpendicular to the velocity.

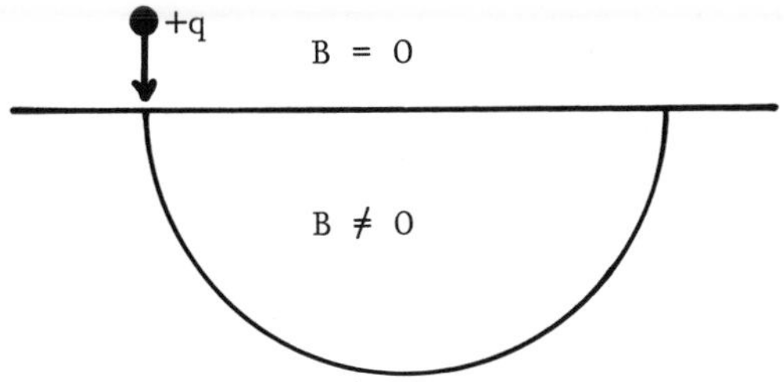

Fig. 6.8

The particle moves with uniform circular motion in the magnetic field. Thus the net force, provided by the magnetic force, must be in toward the center of the circle. For example when the particle enters the field, moving down, the direction of the force at that instant is to the right, in toward the center of the circle. If you point the thumb of your right hand down, the palm of your right hand will be ready to push to the right if your fingers point into the page. Thus, the magnetic field must be into the page.

Problems

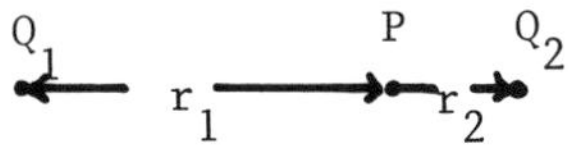

Fig. 6.9

1. In Fig. 6.9, $Q_1 = +2 \times 10^{-6}$ C, $Q_2 = -10^{-6}$ C $r_1 = 3 \times 10^{-2}$ m and $r_2 = 10^{-2}$ m. The electric field at P is

   (a) 7×10^7 N C^{-1} to the right

   (b) 7×10^7 N C^{-1} to the left

   (c) 11×10^7 N C^{-1} to the right

   (d) 90/8 N C^{-1} to the right

2. In Problem 1, a charge $q = +10^{-6}$ C is placed at P. The force exerted on q is

   (a) 70 N to the right

   (b) 70 N to the left

   (c) 110 N to the right

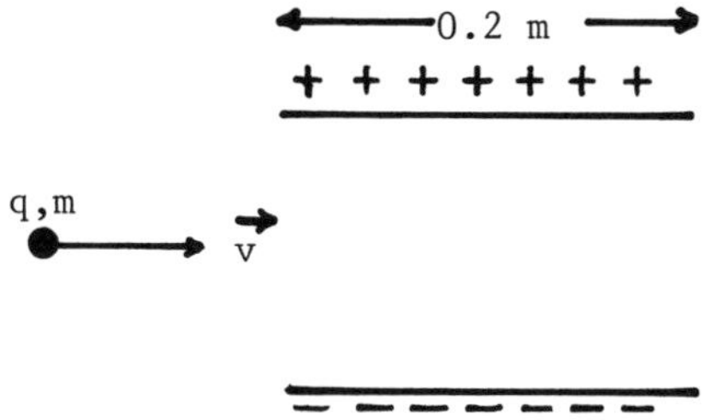

Fig. 6.10

3. A positive charge $q = 10^{-10}$ C with mass $m = 10^{-16}$ kg, moving to the right with velocity $v = 10^5$ m s^{-1}, enters the constant electric field $E = 10^5$ N C^{-1}. The force exerted on the charge by the electric field is

   (a) 10^{-5} N down

   (b) 10^{-5} N up

   (c) 1 N down

   (d) 1 N up

4. In Problem 3, the time for the charge to pass through the length of the plates is

   (a) 1×10^{-6} s

   (b) 2×10^{-6} s

   (c) 3×10^{-6} s

5. In Problem 3, the speed of the particle when it leaves the plates is

   (a) 10^5 m s^{-1}

   (b) $\sqrt{5} \times 10^5$ m s^{-1}

   (c) 3×10^5 m s^{-1}

6. The magnetic field (direction and magnitude) that allows the particle in Problem 3 to pass through the plates with

no deflection is

(a) 10^{-5} N(A-m)$^{-1}$ up

(b) 10^{-5} N(A-m)$^{-1}$ into the page

(c) 1 N(A-m)$^{-1}$ up

(d) 1 N(A-m)$^{-1}$ into the page

(e) 10^{5} N(A-m)$^{-1}$ up

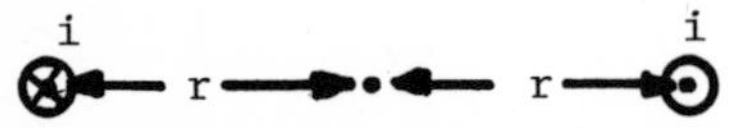

Fig. 6.11

7. Two wires carry equal currents in opposite directions (Fig. 6.11). The magnetic field due to each wire a distance r away is 2 N(A-m)$^{-1}$. The resultant magnetic field due to both wires at P is

(a) Zero

(b) 4 N(A-m)$^{-1}$, up

(c) 4 N(A-m)$^{-1}$, down

(d) 4 N(A-m)$^{-1}$, to the right

8. A particle with charge q and mass m, traveling with velocity v, enters a uniform magnetic field B that is perpendicular to its velocity. The radius of its circular path is r. Another particle with charge 4q and mass 8m, travels with velocity v in the same magnetic field. The radius r' of this particle is

(a) r

(b) 2r

(c) 4r

(d) 8r

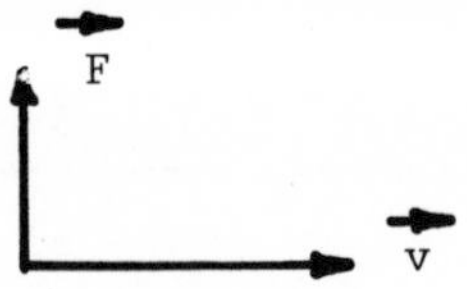

Fig. 6.12

9. A particle with charge $q = 10^{-6}$ C moves to the right with velocity $v = 10^{6}$ m s^{-1} in a uniform magnetic field. The force on the particle F = 0.6 N is up (Fig. 6.12). Which of the following cannot be true?

(a) B = 0.6 N(A-m)$^{-1}$ into the page

(b) B = 1.0 N(A-m)$^{-1}$ in a plane perpendicular to the page at an angle of 37° with v

(c) B = 0.6 N(A-m)$^{-1}$ in the plane of the page

10. The magnetic field along the axis of a solenoid is B = 2 N(A-m)$^{-1}$. A particle of charge $q = 10^{-6}$ C moves with a velocity $v = 10^{5}$ m s^{-1} parallel to the axis. The force on the particle is

(a) Zero

(b) 0.2 N

(c) $0.2\sqrt{2}$ N

Sample Quiz for Chapters 4 through 6

In all cases take $g=10\ \text{m s}^{-2}$.

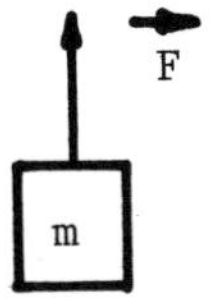

Fig. 1

1. In Fig. 1, the object of mass m = 10 kg is raised with a constant velocity of $10\ \text{m s}^{-1}$. The applied force F is

   (a) Zero

   (b) 10 N

   (c) 100 N

2. For the situation described in Problem 1, the force necessary to give the object an acceleration of $10\ \text{m s}^{-1}$ is

   (a) 10 N

   (b) 100 N

   (c) 110 N

   (d) 200 N

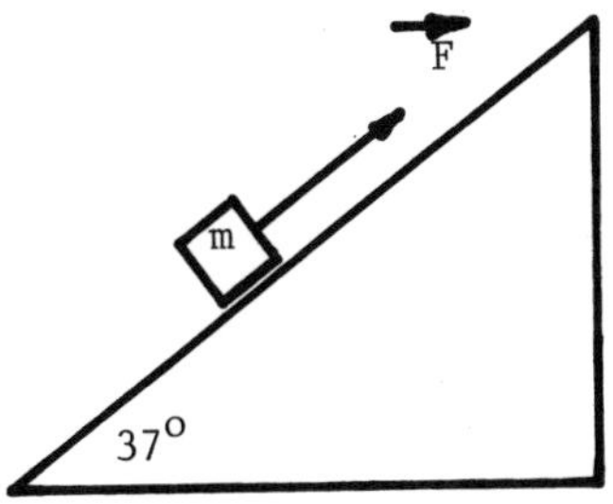

Fig. 2

3. An object of mass m = 2 kg is pulled up an inclined plane with a force F = 20 N (Fig. 2). The acceleration of the object is

   (a) $2\ \text{m s}^{-2}$

   (b) $4\ \text{m s}^{-2}$

   (c) $10\ \text{m s}^{-2}$

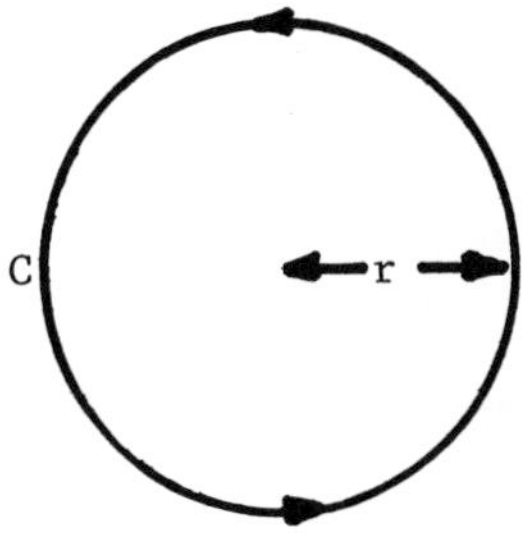

Fig. 3

4. A Ferris wheel of radius r = 12 m makes one revolution in 4π seconds. A person of mass m = 50 kg sits on a seat of the Ferris wheel. The speed of the person is

   (a) $1/4\pi\ \ \text{m s}^{-1}$

   (b) $4\pi\ \ \text{m s}^{-1}$

   (c) $6.0\ \text{m s}^{-1}$

5. The magnitude of the force of the seat on the person at position C in Fig. 3 is

   (a) 150 N

   (b) $50\sqrt{109}$ N

   (c) 500 N

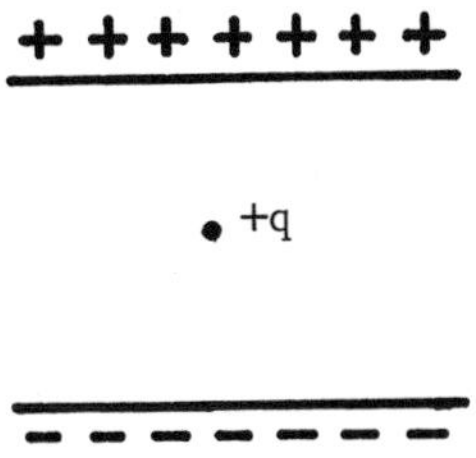

Fig. 4

6. Two parallel plates have total charges +Q and -Q. A particle of charge +q is between the plates (Fig. 4). It experiences a force equal to

(a) Zero

(b) $k_e qQ/d^2$

(c) $2k_e qQ/d^2$

(d) none of the above

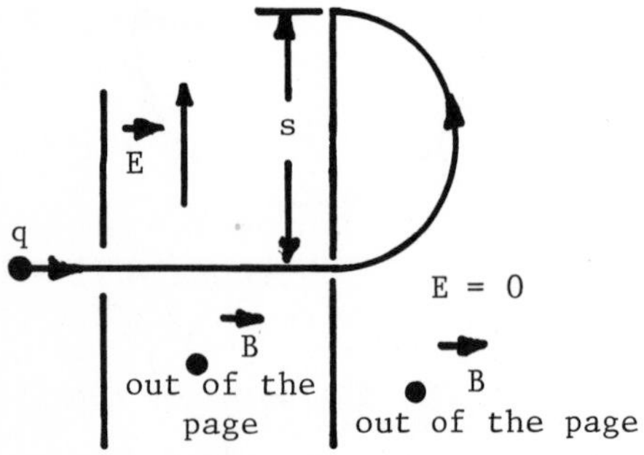

Fig. 5

7. An ion whose charge is known to have magnitude q is observed to move in a straight line through a region in which a magnetic field B points out of the page and an electric field E points up, as shown in Fig. 5. Entering a second region, in which only the magnetic field is present, the ion moves in a semicircular arc as shown. The ion has

(a) a positive charge

(b) a negative charge

8. In Problem 7, the mass of the particle is

(a) $B^2qs/2E$

(b) B^2qs/E

(c) $B\ qs/2$

(d) none of the above

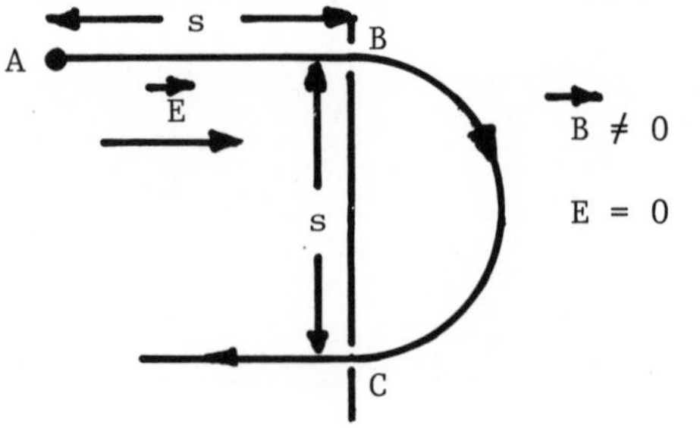

Fig. 6

9. A particle of positive charge q and mass m is released from rest at position A in Fig. 6. It passes first through a region of width s in which there is an electric field E to the right. Then it passes through a slit at B and enters a region in which there is a uniform magnetic field B (there is no electric field), emerging at position C as shown in the figure after traveling in a semicircle, whose diameter is also s. The magnetic field (direction and magnitude) is

(a) $\sqrt{Em/2qs}$, out of the page

(b) $4Em/qs$, out of the page

(c) $2\sqrt{2Em/qs}$, out of the page

(d) $\sqrt{Em/2qs}$, into the page

10. In Problem 9, the total time required for travel from A to C is

(a) $\sqrt{2ms/qE}\ (1 + \pi/4)$

(b) $\sqrt{2ms/qE}$

(c) $5/4\ \sqrt{2ms/qE}$

CHAPTER 7. Conservation Laws

Objectives

You should learn

1. that momentum is a vector quantity and that it is conserved in all collisions and in a closed system
2. that kinetic energy is a scalar quantity and that it is conserved only in elastic collisions
3. the definition of work as the distance moved times the component of the force in the direction of the displacement
4. that the total work done on an object may go into an increase in potential, kinetic, heat and/or mass energy
5. that when no work is done on a system, the sum of the change in potential energy, the change in kinetic energy, the change in heat energy, and the change in mass energy equals zero
6. the expressions for gravitational potential energy, electric potential energy, and elastic potential energy
7. the difference between potential difference and potential energy difference
8. the difference between force and pressure
9. the dependence of temperature on the average kinetic energy of the molecules
10. the definition of angular momentum
11. how to apply the conservation laws to solve problems

Trouble Spots

The conservation of a quantity means the value of the quantity (and its direction, if it is a vector quantity) for a collection of particles remains the same at all times. For example, in the experiment described in the text when a ball of mass m moves with velocity $\vec{v}$ to the right toward six other balls of equal mass, the momentum of the system before is

$$m\vec{v} \text{ to the right.}$$

After the collision, the initial ball comes to rest and the ball farthest to the right goes off with momentum

$$m\vec{v} \text{ to the right.}$$

If you are standing in a boat and the boat is at rest, the initial momentum of the person-boat system is zero. When you jump toward the shore, the boat goes off with momentum that is equal and opposite to your momentum. The final momentum of the system remains equal to zero.

If you hold a ball, the momentum of the ball-earth system is zero. When you release the ball, the earth must move slightly upward throughout the ball's descent to keep the momentum of the system equal to zero.

While momentum is always conserved in a collision, kinetic energy is only conserved in an elastic collision. In all other collisions some of the energy appears as heat. Energy is not a vector quantity. In a closed system, the potential energy may decrease as the kinetic energy increases, but neither has a directional property.

In a closed system, the change in potential energy, rather than the absolute value of potential energy, is the significant quantity. It is customary to assign zero potential energy to:

1. a position on the earth's surface for gravitational potential energy near the earth's surface

2. a large distance from the proton for electric potential energy of the hydrogen atom

3. the equilibrium position for simple harmonic motion

There is an increase in potential energy if you must do work to move the object to that position. For example, the electric potential energy for two unlike charges increases with distance of separation, because you must do work to separate them. Although the force decreases with distance of separation, you must do additional work as you move them farther apart. The work done decreases with distance of separation, but you must add this to the work already done to get that separation. The electric potential energy for two like charges increases with a decrease in separation because you must do work to bring them together.

Potential difference V_{ab} is defined as the potential energy difference $(PE_a - PE_b)$ divided by the charge q:

$$V_{ab} = \frac{PE_a - PE_b}{q}$$

If you are given V_{ab}, you may find the change in potential energy difference by multiplying V_{ab} by q, that is

$$PE_a - PE_b = qV_{ab}$$

The pressure at a point in a liquid at rest is the same in all directions. If it were not, the liquid at that point would move. The pressure in a fluid is independent of the shape of the container. It depends only on the depth of the fluid.

When a particle of mass m moves in a circle of radius r with speed v, its angular momentum L = mvr. If the path is not circular,

$$L = mv_{\perp} r$$

where $v_{\perp}$ is the component of the velocity that is perpendicular to the distance from the axis of rotation. In a closed system, the angular momentum of a system remains constant. When the particle moves in a path for which the distance from the axis of rotation changes, the kinetic energy and the potential energy of the particle changes, but the total energy and angular momentum of the particle remains constant.

Illustrative Examples

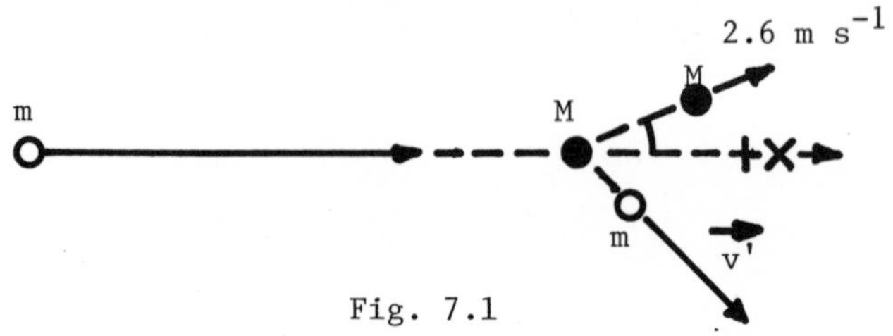

Fig. 7.1

1. A wooden disk of mass m = 1 kg, sliding toward the +x-direction with a velocity of 17 m s^{-1}, collides with a metal disk of mass M = 5 kg, that is initially at rest (Fig. 7.1). After the collision the metal disk moves with a velocity of 2.6 m s^{-1} at an angle of 22.5° above the +x-axis. (a) What is the velocity v' (direction and magnitude) of the wooden disk after the collision? (b) Is this an elastic collision?

(a) The momentum of the metal disk after the collision is

(p_M) = M x 2.6 m s^{-1}, at an angle of 22.5°

= 5 kg x 2.6 m s^{-1}

= 13 kg-m s^{-1}, at an angle of 22.5° with the +x-axis

The x-component of the momentum of the metal disk is

$(p_M)_x = p_M \cos 22.5^\circ$

= 13 kg-m s^{-1} x 12/13

= 12 kg-m s^{-1}

The y-component of the momentum of the metal disk is

$$(p_M)_y = p_M \sin 22.5^o$$
$$= 13 \text{ kg-m s}^{-1} \times 5/13$$
$$= 5 \text{ kg-m s}^{-1}$$

Before the collision, the system had momentum in the x-direction of

$$p_m = m \times 17 \text{ m s}^{-1}$$
$$= 1 \text{ kg} \times 17 \text{ m s}^{-1}$$
$$= 17 \text{ kg-m s}^{-1}$$

and a momentum in the y-direction of zero. From conservation of momentum,

For x-direction

Momentum before = Momentum after

$$17 \text{ kg-m s}^{-1} = 12 \text{ kg-m s}^{-1} + 1 \text{ kg}(v')_x$$
$$5 \text{ m s}^{-1} = (v')_x$$

For y-direction

Momentum before = Momentum after

$$0 = 5 \text{ kg-m s}^{-1} + 1 \text{ kg}(v')_y$$
$$-5 \text{ m s}^{-1} = (v')_y$$

After the collision, the velocity of the wooden disk is

$$v' = \sqrt{(v')_x^2 + (v')_y^2}$$
$$= \sqrt{(5)^2 + (5)^2} \text{ m s}^{-1}$$
$$= 5\sqrt{2} \text{ m s}^{-1}$$

at an angle of 45° below the +x-axis.

(b) The kinetic energy of the wooden disk before the collision is

$$1/2 \text{ (1 kg) } (17 \text{ m s}^{-1})^2 = 144.5 \text{ J}$$

The kinetic energy of the wooden disk after the collision is

$$1/2 \text{ (1 kg) } (5\sqrt{2} \text{ m s}^{-1})^2 = 25.0 \text{ J}$$

The kinetic energy of the metal disk after the collision is

$$1/2 \text{ (5 kg) } (2.6 \text{ m s}^{-1})^2 = 16.9 \text{ J}$$

Since 144.5 J $\neq$ (25.0 + 16.9)J, this is not an elastic collision.

2. The length of an incline is 10 m and its height h = 5 m. How much work must be exerted to move an object of mass m = 2 kg the entire length s of the incline (a) at constant speed (b) changing the speed of the object from 0 to 5 m s^{-1} (c) changing the speed of the object from 0 to 5 m s^{-1}, if the plane exerts a frictional force of 3 N. For (a) and (b) assume that the plane is frictionless.

(a) When the object is moved up the plane with a constant velocity, all the work goes into an increase in potential energy. (Some work must be done, of course, to put the object in motion.)

$$\text{Work done} = \Delta \text{ PE}$$
$$= mgh$$
$$= 2 \text{ kg} \times 10 \text{ m s}^{-2} \times 5 \text{ m}$$
$$= 100 \text{ J}$$

(b) Now work must be done to increase its potential energy and its kinetic energy

$$\text{Work done} = \Delta \text{ PE} + \Delta \text{ KE}$$
$$= 100 \text{ J} + 1/2 \times 2 \text{ kg} \times (5 \text{ m s}^{-1})^2$$
$$= 100 \text{ J} + 25 \text{ J}$$
$$= 125 \text{ J}$$

(c) The frictional force acts along the entire length of the plane or 10 m. The work done against friction equals

$$W_f = F_f \, s$$
$$= 3 \text{ N} \times 10 \text{ m}$$

$= 30$ J

The total work done $= \Delta PE + \Delta KE + W_f$

$= 100$ J $+ 25$ J $+ 30$ J

$= 155$ J

3. A negatively charged particle with $q = -10^{-6}$ C is initially a distance $r_1 = 10^{-2}$ m from a positively charged particle with $q' = +10^{-6}$ C. (a) How much work must be done to separate the charges by a distance $r_2 = 2 \times 10^{-2}$ m, if they are moved apart without a change in velocity? (b) After the charges are separated by a distance of 2×10^{-2} m, they are released. What is the change in the kinetic energy of the system when their distance of separation is 1×10^{-2} m?

(a) Since the charges are separated at constant speed, the total work done goes into an increase in potential energy.

$$W_{1 \rightarrow 2} = PE_2 - PE_1$$

$$= \frac{-k_e qq'}{r_2} - \frac{-k_e qq'}{r_1}$$

$$= k_e qq'(1/r_1 - 1/r_2)$$

$$= k_e qq'(10^2 - 10^2/2) \text{ m}^{-1}$$

$$= k_e qq' \ (10^2/2) \text{ m}^{-1}$$

$$= 9 \times 10^9 \text{ N-m C}^{-2}(10^{-12} \text{ C}^2)(10^2/2)$$

$$= 0.45 \text{ N-m} = 0.45 \text{ J}$$

(b) $PE_1 + KE_1 = PE_2 + KE_2$

$KE_1 - KE_2 = PE_2 - PE_1$

$= 0.45$ J

4. The equation of motion for an object of mass $m = 1$ kg attached to a spring of spring constant k is

$$s = 0.13 \text{ m} \cos \pi t/2$$

Find (a) the amplitude A, (b) the period T of the motion, (c) the spring constant k, (d) the displacement s of the object at $t = 1/4$ s, (e) the acceleration of the object at $t = 1/4$ s, and (f) the velocity v of the object at $t = 1/4$ s.

In general, for simple harmonic motion

$$s = A \cos 2\pi t/T$$

(a) By comparison we see that $A = 0.13$ m

(b) and $\pi t/2 = 2\pi t/T$, or $T = 4$ s

(c) The period of the spring

$$T = 2\pi\sqrt{m/k}$$

or

$$k = 4\pi^2 m/\ T^2$$

$$= 4\pi^2(1 \text{ kg})/(16 \text{ s}^2)$$

$$= \pi^2/4 \text{ N-m}^{-1}$$

(d) At $t = 1/4$ s

$$s = 0.13 \text{ m} \cos \pi/8$$

Since $\pi/8 = 22.5^\circ$

$$s = 0.13 \text{ m} \cos 22.5^\circ$$

$$= 0.13 \text{ m} \times 12/13$$

$$= 0.12 \text{ m}$$

(e) $a = -4\pi^2 s/T^2$

$$= -4\pi^2(0.12 \text{ m})/(16 \text{ s}^2)$$

$$= -0.03\ \pi^2 \text{ m s}^{-2}$$

(f) When the mass has its maximum displacement of 0.13 m, the object is at rest and its total energy equals its potential energy. In general, $PE = 1/2\ ks^2$. For $s = A$,

$$E = PE + KE$$

$= 1/2\ kA^2 + 0$

$= 1/2\ (\pi^2/4\ N\text{-}m^{-1})(0.0169\ m^2)$

$= 0.0169\ \pi^2/8\ J$

When s= 0.12 m,

$PE = 1/2\ (\pi^2/4\ N\text{-}m^{-1})(0.0144\ m^2)$

$= 0.0144\ \pi^2/8\ J$

From conservation of energy,

$0.0169\ \pi^2/8\ J = 0.0144\ \pi^2/8\ J + 1/2(1\ kg)v^2$

$0.0025\ \pi^2/8\ J = 1/2 \times 1\ kg \times v^2$

$\pi(0.025)m\ s^{-1} = v$

5. Helium gas follows the ideal-gas law quite closely. Imagine that $10^{-3}\ m^3$ of helium is initially at a pressure of $10^5\ N\ m^{-2}$ and a temperature of 300° K. Find (a) the volume if it is cooled to 10° K at constant pressure (b) the pressure if it is allowed to expand back to its original volume at 10° K.

(a) In general

$P_2V_2/T_2 = P_1V_1/T_1$

For $P_2 = P_1$,

$V_2 = V_1T_2/T_1$

$$V_2 = \frac{(10^{-3}\ m^{-3})(10^o\ K)}{300^o\ K}$$

$= 3.3 \times 10^{-5}\ m^3$

(b) Now

$P_3 = P_2V_2/V_3$

$$P_3 = \frac{(10^5\ N\ m^{-2})(3.3 \times 10^{-5}\ m^3)}{10^{-3}\ m^3}$$

$= 3.3 \times 10^{-3}\ N\ m^{-2}$

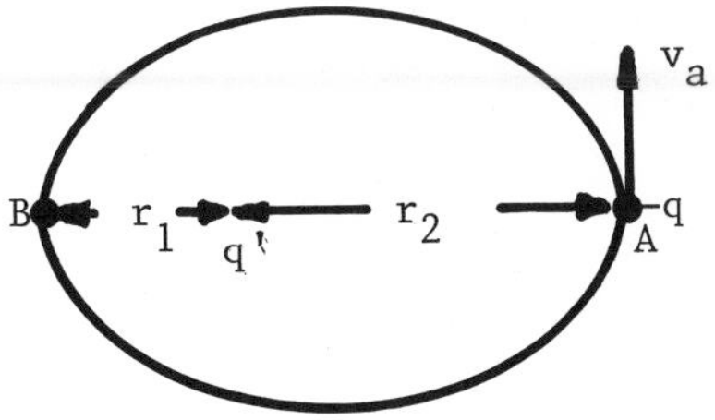

Fig. 7.2

6. A negatively charged particle of charge $q = -10^{-6}$ C and mass $m = 10^{-10}$ kg travels in an elliptical orbit around a positive charge $q' = +10^{-6}$ C. When it is at A, the distance of $-q$ from q' is $r_1 = 6 \times 10^{-5}$ m. When it is at B, the distance of $-q$ from q' is $r_2 = 3 \times 10^{-5}$ m. Find (a) the ratio of the particle's velocity at B to its velocity at A. (b) Use conservation of energy to find v_a and v_b.

(a) At A and B, the velocity of the particle is perpendicular to the line from the axis of rotation. From conservation of angular momentum,

$L_A = L_B$

$mv_ar_1 = mv_br_2$

$v_b/v_a = r_1/r_2$

$= 6/3$

$= 2$

(b) Total energy at A = Total energy at B

$PE_A + KE_A = PE_B + KE_B$

$$\frac{-k_eqq'}{r_1} + 1/2\ mv_a^2 = \frac{-k_eqq'}{r_2} + 1/2\ mv_b^2$$

$$\frac{-k_eqq'}{r_1} + \frac{k_eqq'}{r_2} = 1/2\ m(v_b^2 - v_a^2)$$

$k_eqq'(1/r_2 - 1/r_1) = 1/2\ m(4v_a^2 - v_a^2)$

$$\frac{2k_e qq'(1/r_2 - 1/r_1)}{3m} = v_a^2$$

$$\frac{2k_e qq'(10^5/3 - 10^5/6)\ m^{-1}}{3m} = v_a^2$$

$$\frac{2k_e qq'(10^5/6)\ m^{-1}}{3m} = v_a^2$$

$$\frac{k_e qq'(10^5)\ m^{-1}}{9m} = v_a^2$$

$$\frac{9 \times 10^9\ N\text{-}m^2\ C^{-2}(10^{-12}C^2)(10^5)m^{-1}}{3 \times 20^{-10}\ kg} = v_a^2$$

$$10^{12}\ m^2\ s^{-2} = v_a^2$$

$$10^6\ m\ s^{-1} = v_a$$

$$v_b = 2\ v_a = 2 \times 10^6\ m\ s^{-1}$$

7. A proton of mass m and charge e is accelerated through a potential difference V_{ab}. The proton is initially at rest and gains a velocity v. If an alpha particle of mass 4m and charge 2e is accelerated through the same potential difference, compare the velocity v' gained by the alpha particle to that of the proton.

In both cases, the change in potential energy equals the increase in kinetic energy.

For the proton,

$$1/2\ mv^2 = eV_{ab}$$

or

$$v = \sqrt{2eV_{ab}/m}$$

For the alpha particle,

$$1/2\ (4m)v'^2 = 2eV_{ab}$$

or

$$v' = \sqrt{eV_{ab}/m}$$

$$\frac{v'}{v} = \frac{\sqrt{eV_{ab}/m}}{\sqrt{2eV_{ab}/m}}$$

$$= 1/\sqrt{2}$$

Problems

1. Particle A and Particle B have the same momentum, but Particle B has one-half the mass of Particle A. The ratio of the kinetic energy B to the kinetic energy of A is

(a) 1

(b) 2

(c) 4

2. A ball of mass $m_1 = 2$ kg traveling to the right with velocity $v_1 = 5\ m\ s^{-1}$ along a level track collides with a ball of mass $m_2 = 1$ kg traveling to the left with velocity $v_2 = 7\ m\ s^{-1}$. After the collision the balls stick together and move off with a common velocity v. This is a one-dimensional problem. The velocity v is

(a) $1\ m\ s^{-1}$ to the right

(b) $2\ m\ s^{-1}$ to the left

(c) $\sqrt{33}\ m\ s^{-1}$ to the right

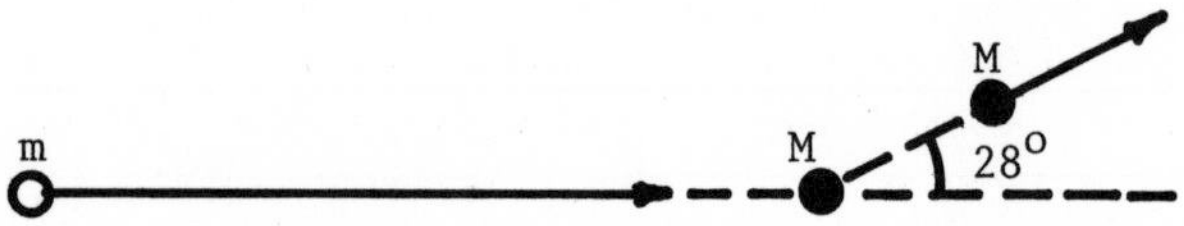

Fig. 7.3

3. A wooden disc of mass m = 1 kg, sliding toward the east with a velocity of $15\ m\ s^{-1}$, collides with a metal disk of mass M = 5 kg that is initially at rest. After the

collision, the metal disk moves off at an angle of 28° north of east with a velocity of 3.4 m s^{-1}, as shown in Fig. 7.3. The velocity of the wooden disk after the collision is

(a) 2 m s^{-1}, east

(b) 8 m s^{-1}, south

(c) 11.6 m s^{-1}, east

4. An electron is accelerated from rest through a potential difference of 180 V. Take the charge and the mass of the electron to be 1.6 x 10^{-19} C and 9 x 10^{-31} kg, respectively. The final velocity of the electron is

(a) 180 m s^{-1}

(b) 8 x 10^{6} m s^{-1}

(c) 32 x 10^{12} m s^{-1}

5. An electron of charge -e and mass m is in a circular orbit of radius r about a proton of charge +e with speed v. The total energy of the system is

(a) $1/2\ mv^2$

(b) $-k_e e^2/r$

(c) $-k_e e^2/2r$

(d) $k_e e^2/2r$

6. The amplitude for a mass m = 2 kg vibrating at the end of a spring is 0.5 m. When the displacement is 0.3 m from the equilibrium position, it has a speed of 0.8 m s^{-1}. The spring constant k in N-m^{-1} is

(a) 5.12

(b) 8.0

(c) 32

7. A submarine is lying on its side in 100 m deep water. The escape hatch door is 1.0 m x 0.5 m. The density of the water is 1.025 x 10^{3} kg m^{-3}. The pressure inside the submarine is atmospheric pressure or 10^{5} N m^{-2}. The force on the door of the submarine is

(a) 4.5 x 10^{5} N

(b) 5.0 x 10^{5} N

(c) 5.5 x 10^{5} N

8. If the average speed of molecules in a gas is tripled, the temperature is increased by a factor of

(a) 3

(b) 6

(c) 9

(d) 18

9. An engine takes in energy from a source at 800° K and rejects heat at a temperature T. The efficiency of the engine is 60%. Temperature T is

(a) 320° K

(b) 500° K

(c) 1333° K

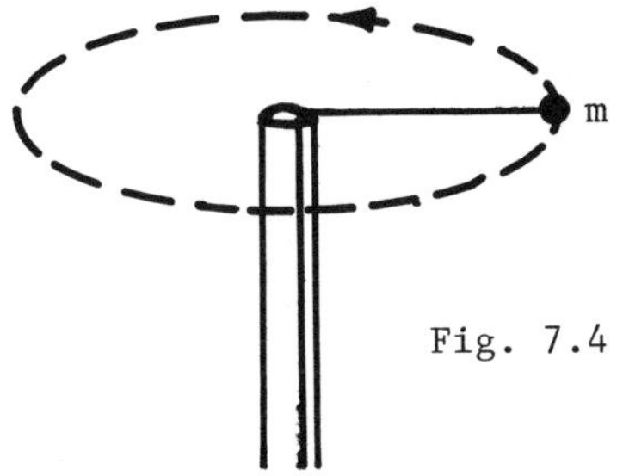

Fig. 7.4

10. An object of mass m = 0.5 kg is at one end of a string. The other end passes through a glass tube, as shown in Fig. 7.4. The mass is twirled in a circle of radius r = 1 m. When the string is pulled down the radius of the circle becomes 0.5 m. If the initial frequency

of the motion is f, the frequency at radius of 0.5 m is

(a) f/4

(b) f/2

(c) f

(d) 2f

(e) 4f

11. In pair annihilation, an electron and a positron (the anti-particle of the electron) disappear and two high energy gamma rays are formed. Which of the following conservation laws are violated in pair annihilation?

(a) conservation of charge

(b) conservation of momentum

(c) conservation of energy

(d) conservation of angular momentum

(e) none of the above

CHAPTER 8. Transfer of Energy by Particles Electric Circuits and Geometric Optics

Objectives

You should learn how

1. to draw circuit diagrams
2. to distinguish between resistors in series and in parallel
3. to calculate the terminal voltage across a seat of emf
4. to use conservation of energy to check the energy dissipated in a circuit
5. to convert a galvanometer to an ammeter or to a voltmeter
6. to distinguish between reflection and refraction
7. to find the critical angle
8. to draw ray diagrams for lenses and for mirrors
9. to find linear and angular magnification
10. to use a virtual object to find the final image for a combination of lenses

Trouble Spots

When resistors are connected in parallel, the resistance of the combination is less than that of any of the individual resistors. Do not make the mistake, however, of forgetting to invert the reciprocal to find the equivalent resistance. For example, imagine that two resistors of resistance 3 ohm and 6 ohm are connected in parallel. The reciprocal of the equivalent resistance is

$$\frac{1}{R} = \frac{1}{3} + \frac{1}{6} \text{ ohm}^{-1}$$

$$= \frac{2+1}{6} \text{ ohm}^{-1}$$

$$= \frac{3}{6} \text{ ohm}^{-1}$$

and

$$R = \frac{6}{3} \text{ ohm}$$

$$= 2 \text{ ohm}$$

The total current does not split in proportion to the resistance of resistors in parallel. For example, if the total current is 1.0 ampere, the current through the 3 ohm and 6 ohm resistors is not, 1/3 A and 1/6 A, respectively. The sum of the current in each resistor must equal the total current. If i_1 is the current through the 3 ohm resistor and i_2 the current through the 6 ohm resistor, the total current

$$1 \text{ A} = i_1 + i_2$$

or $$1 \text{ A} - i_1 = i_2$$

Since the potential difference across resistors in parallel is the same,

$$i_1 \times 3 \text{ ohm} = i_2 \times 6 \text{ ohm}$$

$$i_1 \times 3 \text{ ohm} = (1 \text{ A} - i_1) \times 6 \text{ ohm}$$

$$3i_1 = 6 \text{ A} - 6i_1$$

$$9i_1 = 6\text{A}$$

$$i_1 = 2/3 \text{ A}$$

$$i_2 = (1 \text{ A} - i_1)$$

$$= (1 \text{ A} - 2/3 \text{ A})$$

$$= 1/3 \text{ A}$$

When a seat of electromotive force with emf = ξ and internal resistance r delivers energy to the rest of the circuit, its terminal voltage

$$V_{ab} = \xi - ir$$

When the seat absorbs energy from the rest of the circuit, its terminal voltage is

$$V_{ab} = \xi + ir$$

The terminal voltage across a seat of emf is never equal to the current i through the cell times the internal resistance of the cell.

An ammeter should never be connected directly across a battery because it has a very low resistance. When measuring low resistances by measuring the current through it and the potential difference across it, you should place the voltmeter in parallel with the resistor and the ammeter in series with this parallel combination. When measuring large resistances, you should place the ammeter in series with the resistor and the voltmeter in parallel across the series combination of the resistor and ammeter.

When light hits a plane mirror, the angle of incidence equals the angle of reflection. When light travels from air to a more dense medium, the angle of incidence is greater than the angle of refraction.

Light rays are reversible. When light travels from a more dense medium to air, the angle of incidence is less than the angle of refraction. When light enters from a more dense medium, it is totally internally reflected if the angle of incidence is greater than the critical angle.

Two of the following three rays from the top of an object may be used to locate the top of the image for a thin converging lens:

1. a ray parallel to the axis of the lens that is refracted through the focal point.

2. a ray through the center of the lens that experiences negligible refraction

3. a ray through the focal point that is refracted parallel to the axis of the lens

If the bottom of the object lies on the axis of the lens, the bottom of the image will also lie on the axis of the lens.

For a diverging lens

1. a ray parallel to the axis of the lens diverges as though it had come from the focal point on the same side of the lens as the object

2. a ray through the center of the lens experiences negligible refraction

3. a ray that would pass through the focal point on the side of the lens opposite to that of the object is diffracted parallel to the axis of the lens

Light rays pass through lenses although a virtual image may be perceived to be on the same side of the lens as the object. Light rays are always reflected from mirrors although a virtual image may appear to be found in back of the mirror.

Illustrative Examples

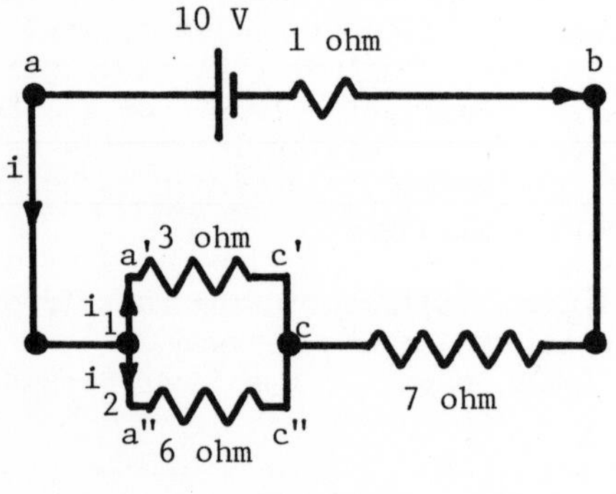

Fig. 8.1

1. For the circuit of Fig. 8.1, find (a) the total resistance of the circuit, (b) the total current i, (c) i_1 and i_2, (d) V_{ac} (e) V_{cb}, and (f) V_{ab}.

(a) $\frac{1}{R_{ac}} = \frac{1}{3} + \frac{1}{6}\ \text{ohm}^{-1}$

$R_{ac} = 2$ ohm

$R = R_{ac} + R_{cb} + R_{ab}$

$= (2 + 7 + 1)$ ohm

$= 10$ ohm

(b) $i = \xi/R$

$= 10\text{ V}/10\text{ ohm}$

$= 1$ A

(c) $V_{a'c'} = V_{a''c''}$

$3i_1 = (1\text{ A} - i_1)6$

$i_1 = 2/3$ A

$i_2 = 1/3$ A

(d) $V_{ac} = iR_{ac}$

$= 1\text{ A} \times 2$ ohm

$= 2$ V

(e) $V_{cb} = iR_{cb}$

$= 1\text{ A} \times 7$ ohm

$= 7$ V

(f) $V_{ab} = \xi - ir$

$= 10\text{ V} - (1\text{ A})(1\text{ ohm})$

$= 10\text{ V} - 1\text{ V}$

$= 9$ V

Note that $V_{ab} = V_{ac} + V_{cb} = V_a - V_c + (V_c - V_b)$

$= 2\ V + 7\ V$

$= 9\ V$

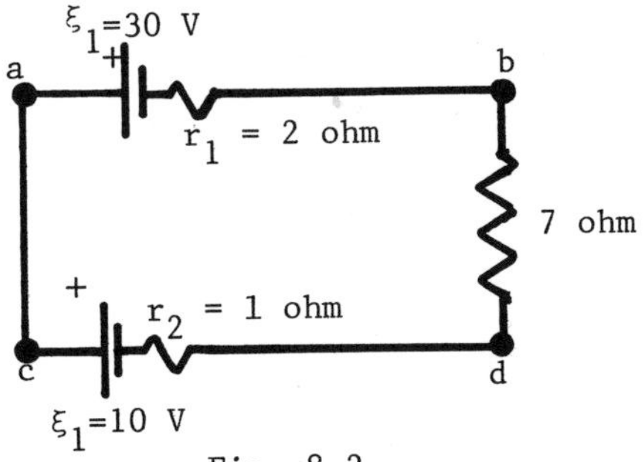

Fig. 8.2

2. For the circuit of Fig. 8.2 find (a) the total resistance (b) the current i, (c) V_{cd}, (d) V_{db}, (e) V_{ab}, (f) Power P_1 delivered by ξ_1, (g) Power P_2 consumed by ξ_2, (h) Power P_3 dissipated in heat.

(a) $R_t = r_2 + R + r_1$

$= (1 + 7 + 2)$ ohm

$= 10$ ohm

(b) $i = \dfrac{\xi_1 - \xi_2}{R}$

$= \dfrac{20\ V}{10\ \text{ohm}}$

$= 2\ A$

(c) $V_{cd} = \xi_2 + ir_2$

$= 10\ V + (2\ A)(1\ \text{ohm})$

$= 10\ V + 2\ V$

$= 12\ V$

(d) $V_{db} = iR$

$= 1\ A \times 7$ ohm

$= 7\ V$

(e) $V_{ab} = \xi_1 - ir_1$

$= 30\ V - (2\ A)(2\ \text{ohm})$

$= 26\ V$

(f) $P_1 = \xi_1 i$

$= (30\ V)(2\ A)$

$= 60\ W$

(g) $P_2 = \xi_2 i$

$= (10\ V)(2\ A)$

$= 20\ W$

(h) $P_3 = i^2 R_t$

$= (4\ A^2)(10\ \text{ohm})$

$= 40\ W$

Note that $P_1 = P_2 + P_3$.

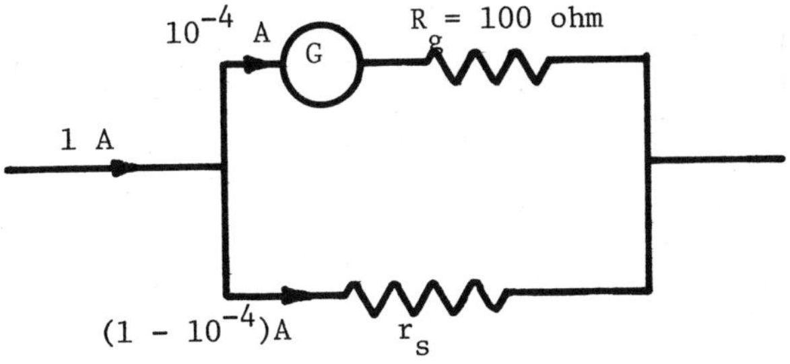

Fig. 8.3

3. A galvanometer, with a coil resistance $R_g = 100$ ohm, has maximum deflection for a current of 10^{-4} A. Show how to convert the galvanometer to an ammeter with a full scale deflection of 1.0 A.

A shunt resistance r_s should be wired in parallel with the galvanometer such that 10^{-4} A goes through the galvanometer and $(1 - 10^{-4})$ A goes through the shunt.

Since R_g and r_s are in parallel, the potential difference across both is the same.

$$(10^{-4}\ A)(100\ ohm) = (1 - 10^{-4})A\ r_s$$

Since $(1 - 10^{-4})$ is approximately equal to 1, we may write

$$(10^{-4}\ A)(100\ ohm) = 1\ A\ r_s$$

and $$10^{-2}\ ohm = r_s$$

The wiring of the galvanometer for conversion to an ammeter is shown in Fig. 8.3.

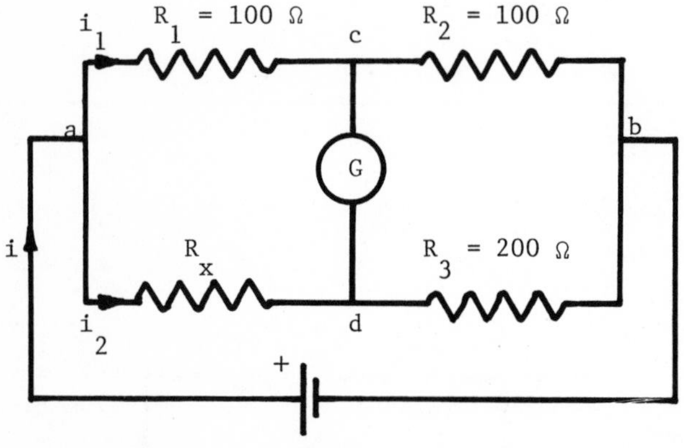

Fig. 8.4

4. In the circuit of Fig. 8.4, the galvanometer G reads zero. Find (a) V_{cd}, (b) i_1, (c) V_{cb}, (d) V_{db}, (e) i_2, (f) V_{ac}, and (g) R_x.

(a) Since no charge flows through the galvanometer,

$$V_c = V_d$$

or $$V_{cd} = V_c - V_d = 0$$

(b)

$$i_1 = \frac{10\ V}{R_1 + R_2}$$

$$= \frac{10\ V}{200\ ohm}$$

$$= 0.05\ A$$

(c) $$V_{cb} = i_1R_2$$

$$= 0.05\ A \times 100\ ohm$$

$$= 5\ V$$

(d) Since $V_c = V_d$, $V_{cb} = V_{db} = 5\ V$

(e) $$i_2 = V_{db}/R_3$$

$$= 5\ V/\ 200\ ohm$$

$$= 0.025\ A$$

(f) $$V_{ac} = i_1R_1$$

$$= 0.05\ A \times 100\ ohm$$

$$= 5\ V$$

(g) Since $V_c = V_d$, $V_{ac} = 5\ V = V_{ad}$

$$V_{ad} = i_2R_x$$

$$5\ V = (0.025\ A)R_x$$

$$200\ ohm = R_x$$

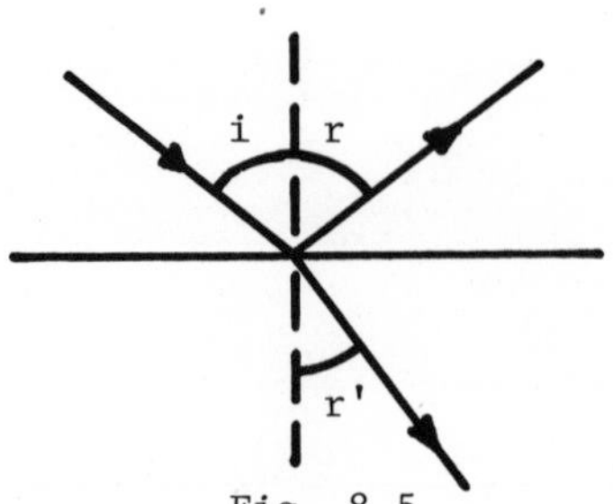

Fig. 8.5

5. Light is incident upon water at an angle of incidence i = 53°. The index of refraction for water is 4/3. Find (a) the angle of reflection r and (b) the angle of refraction r' (Fig. 8.5).

(a) The angle of reflection equals the angle of incidence = 53°.

(b) $$\frac{\sin i}{\sin r'} = n$$

$$\sin r' = \frac{\sin i}{n}$$

$$= \frac{\sin 53^o}{n}$$

$$= \frac{4/5}{4/3}$$

$$= 3/5$$

$$r' = 37^o$$

6. In Problem 5 if the light ray is incident from water to air at an angle of 53°, what will happen to it?

For this case,

$$\frac{\sin 53^o}{\sin \theta} = \frac{1}{n}$$

$$\sin \theta = n \sin 53^0$$

$$= 4/3 \times 4/5$$

$$= 16/15$$

Since the sine of an angle cannot be greater than one, the ray is totally internally reflected.

7. An object is 20 cm to the right of a converging lens of focal length f = 15 cm. Find (a) by calculation, the position of the object and its magnification (b) by a ray diagram, the position of the object.

(a) $$\frac{1}{p} + \frac{1}{q} = \frac{1}{f}$$

$$\frac{1}{q} = \frac{1}{f} - \frac{1}{p}$$

$$= \frac{1}{15} - \frac{1}{20} \text{ cm}^{-1}$$

$$= \frac{4 - 3}{60} \text{ cm}^{-1}$$

$$q = 60 \text{ cm}$$

$$m = \frac{q}{p}$$

$$= \frac{60 \text{ cm}}{20 \text{ cm}}$$

$$= 3$$

(b) The ray diagram is shown in Fig. 8.6 on page 57.

8. Repeat Problem 7 for a diverging lens with focal length f = -15 cm.

(a) $$\frac{1}{p} + \frac{1}{q} = \frac{1}{f}$$

$$\frac{1}{q} = \frac{1}{f} - \frac{1}{p}$$

$$= \frac{-1}{15} - \frac{1}{20} \text{ cm}^{-1}$$

$$= \frac{-4 - 3}{60} \text{ cm}^{-1}$$

$$q = -60/7 \text{ cm} = -8.57 \text{ cm}$$

$$m = q/p$$

$$= -8.57/20$$

$$= -0.428$$

When the magnification is negative, the image is erect. When it is positive, the image is inverted.

(b) The ray diagram is shown in Fig. 8.7 on page 57.

9. An object is 9 cm from a convex mirror of focal length f = 5 cm. Find the position of the image (a) by calculation (b) by a ray diagram.

(a) $$\frac{1}{p} + \frac{1}{q} = \frac{1}{f}$$

$$\frac{1}{q} = \frac{1}{f} - \frac{1}{p}$$

$$= \frac{1}{5} - \frac{1}{9} \text{ cm}^{-1}$$

$$= \frac{9 - 5}{45} \text{ cm}^{-1}$$

$$q = 11.25 \text{ cm}$$

(b) The ray diagram is shown in Fig. 8.8 on page 57.

10. Repeat Problem 9 for a convex mirror of focal length f = - 5 cm.

(a) $\frac{1}{p} + \frac{1}{q} = \frac{1}{f}$

$$\frac{1}{q} = \frac{1}{f} - \frac{1}{p}$$

$$= \frac{-1}{5} - \frac{1}{9}$$

$$= \frac{-9 - 5}{45}$$

$$q = -3.21 \text{ cm}$$

(b) The ray diagram is shown in Fig. 8.9 on page 57.

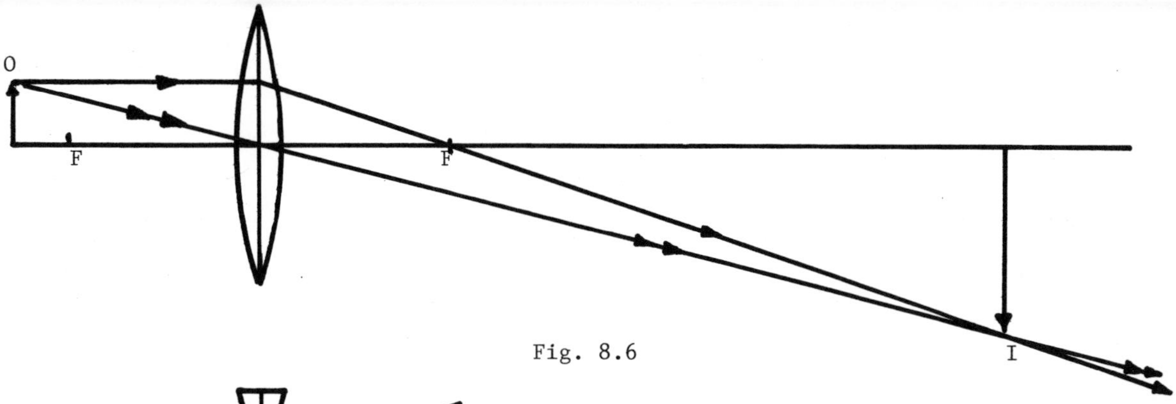

Fig. 8.6

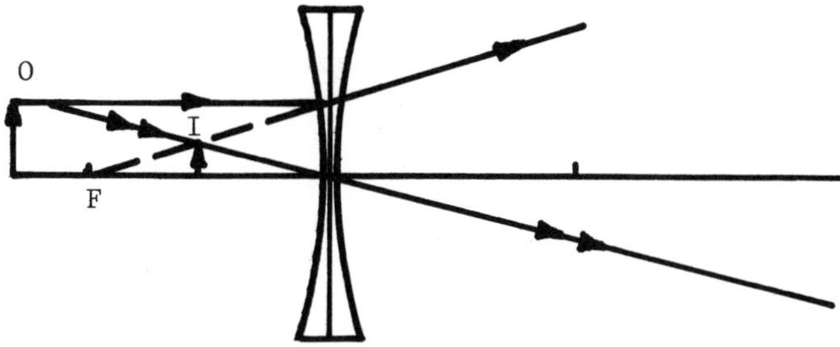

Fig. 8.7

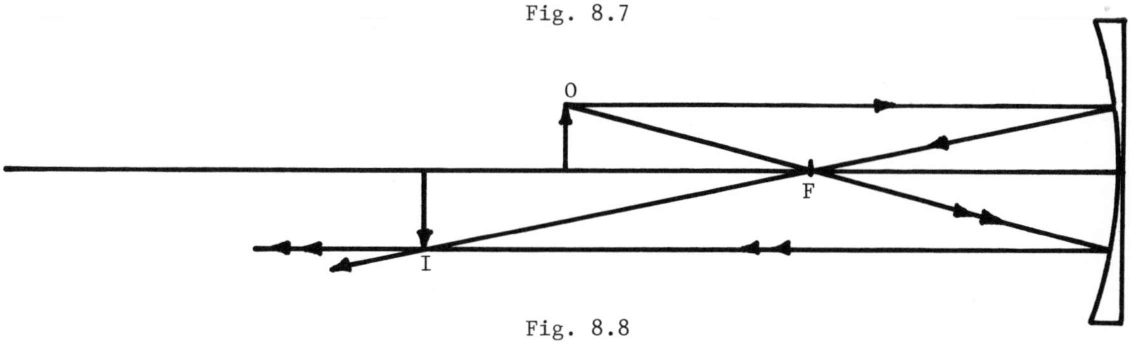

Fig. 8.8

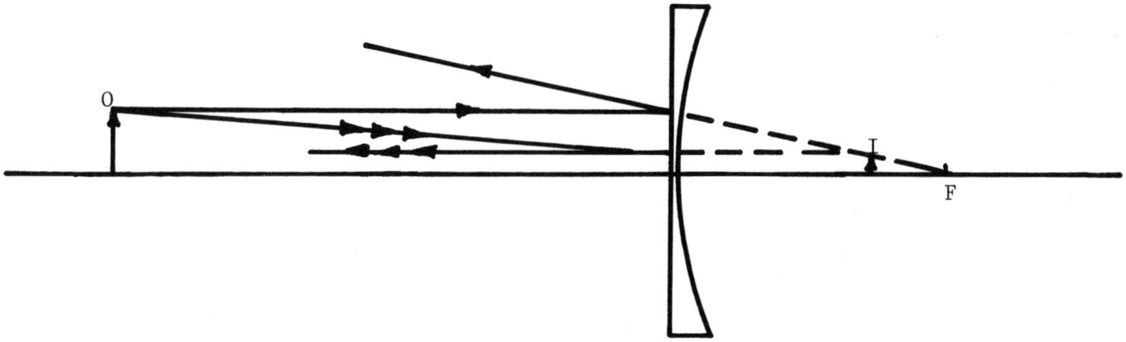

Fig. 8.9

Problems

1. Three resistors of resistance 2 ohm, 4 ohm, and 12 ohm are connected in series. The equivalent resistance of the three resistors is

(a) 1.2 ohm

(b) 18 ohm

2. Three resistors of resistance 2 ohm, 4 ohm, and 12 ohm are connected in parallel. The equivalent resistance of the three resistors is

(a) 5/6 ohm

(b) 1.2 ohm

(c) 18 ohm

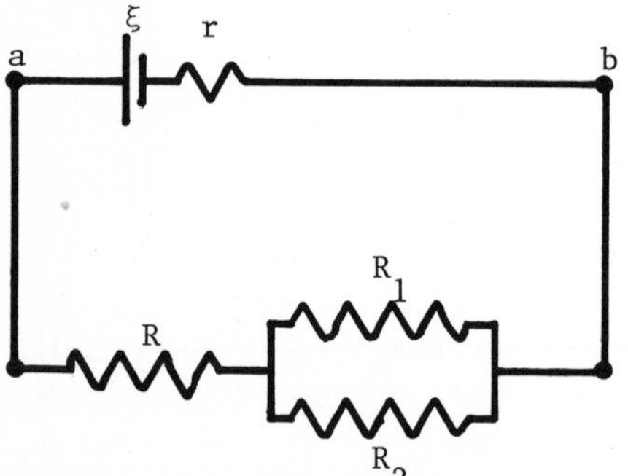

Fig. 8.10

3. For the circuit of Fig. 8.10, the battery has an emf ξ = 18 V and an internal resistance r = 1 ohm. The resistance of the resistors is R = 6 ohm, R_1 = 3 ohm, and R_2 = 6 ohm. The current in R is

(a) 9/8 A

(b) 2 A

(c) 3 A

4. In Problem 3, the current in R_1 is

(a) 2/3 A

(b) 4/3 A

(c) 6 A

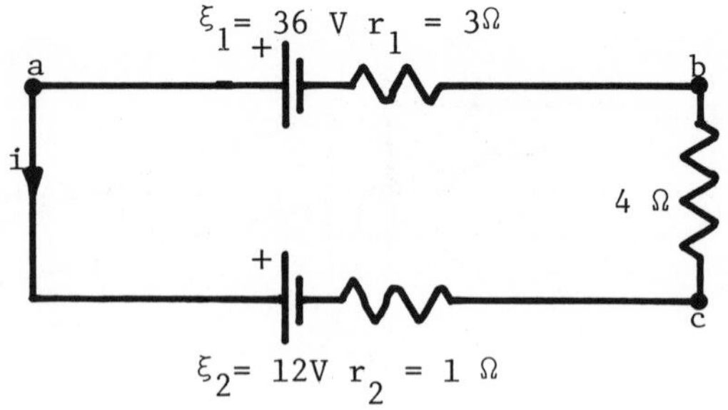

Fig. 8.11

5. For the circuit of Fig. 8.11, i is

(a) 3 A

(b) 4 A

(c) 6 A

6. For the circuit of Fig. 8.11, V_{ac} is

(a) 6 V

(b) 9 V

(c) 12 V

7. A galvanometer has a resistance of 500 ohm and registers full scale deflection for a current of 10^{-3} A. To convert it to a voltmeter with a full scale reading of 1.5 V, you should connect a resistor of

(a) 500 ohm

(b) 1000 ohm

(c) 1500 ohm

This resistor should be connected

(d) in parallel with the galvanometer

(e) in series with the galvanometer

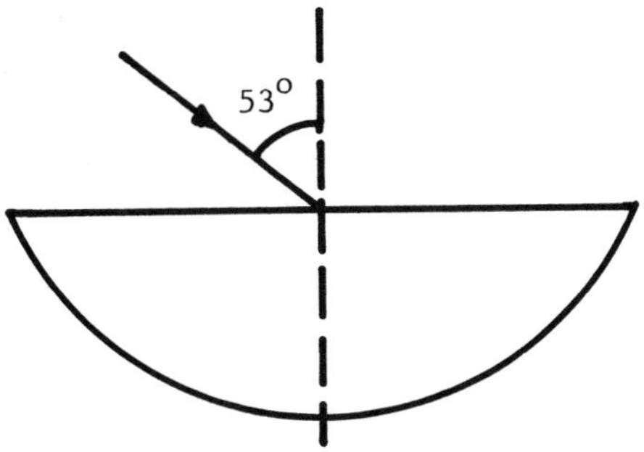

Fig. 8.12

8. A light ray is incident upon a semicircular container of water at an angle of 53° as shown in Fig. 8.12. The index of water is 4/3. Trace the path of the ray through the water and out into the air.

9. An object is 30 cm from a converging lens of focal length f = 40 cm. The image distance is

   (a) 17.14 cm

   (b) 40 cm

   (c) 120 cm

10. Which of the following best describes the image in Problem 9?

   (a) real, inverted, magnified

   (b) real, erect, diminished

   (c) virtual, erect, diminished

11. An object 1 cm high is 100 cm from a converging lens. The image is 3 cm high. The focal length of the lens is

   (a) 75 cm

   (b) 100 cm

   (c) 300 cm

12. An object is 40 cm to the left of a converging lens of focal length f = 30 cm. A diverging lens of focal length f' = -20 cm is 20 cm to the right of the converging lens. The final image for this lens combination is

   (a) 5 cm to the left of the converging lens

   (b) 16.6 to the left of the diverging lens

   (c) 100 cm to the right of the diverging lens

Sample Quiz for Chapters 7 and 8

Take $g = 10\ m\ s^{-2}$

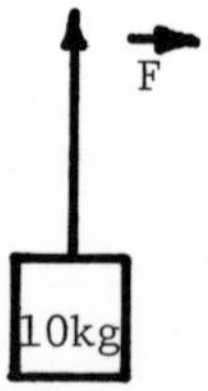

Fig. 1

1. A person lifts a 10 kg-object with a constant force of 200 N. If the object is initially at rest, its velocity at a displacement of 5 m is

   (a) 0

   (b) $10\ m\ s^{-1}$

   (c) $10\sqrt{2}\ m\ s^{-1}$

2. An object of mass m = 0.5 kg attached to a spring with spring constant $k = 8\ N\ m^{-1}$ has a velocity of zero when it is displaced 0.5 m from its equilibrium position. Its velocity when it is 0.3 m from its equilibrium position is

   (a) $1.6\ m\ s^{-1}$

   (b) $2.0\ m\ s^{-1}$

   (c) $2.56\ m\ s^{-1}$

3. An electron of charge $e = 1.6 \times 10^{-19}$ C and mass $m = 9.1 \times 10^{-31}$ kg is accelerated through a potential difference $V_{ab} = 220$ V. It then enters a magnetic field $B = 10^{-4}\ N(A\text{-}m)^{-1}$ that is perpendicular to its velocity. The radius of the electron's circular path is

   (a) 0.25 m

   (b) 0.50 m

   (c) 1.0 m

4. The temperature of a volume of gas is 300° K. When the temperature of the gas is raised to 600° K, the average velocity of the gas molecules is increased by a factor of

   (a) $\sqrt{2}$

   (b) 2

   (c) 3

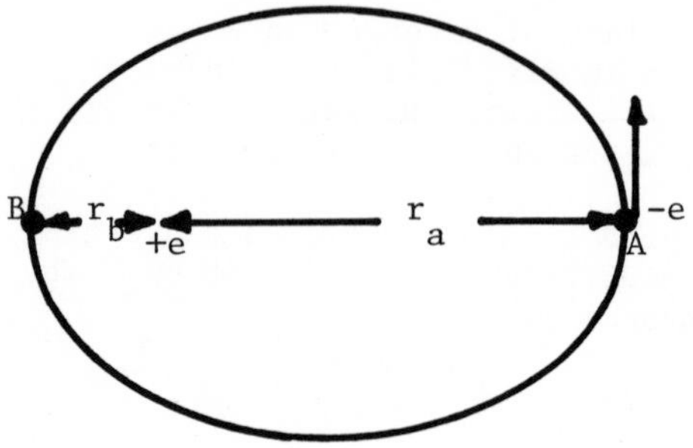

Fig. 2

5. An electron of charge -e is in an elliptical orbit about a proton of charge +e. At position a, its angular momentum is L_a and its potential energy is PE_a (Fig. 2). At position b, its angular momentum is L_b and its potential energy is PE_b. Which of the following statements is true?

   (a) $L_a > L_b$, $PE_a = PE_b$

   (b) $L_a = L_b$, $PE_a = PE_b$

   (c) $L_a = L_b$, $PE_a > PE_b$

   (d) $L_a = L_b$, $PE_b > PE_a$

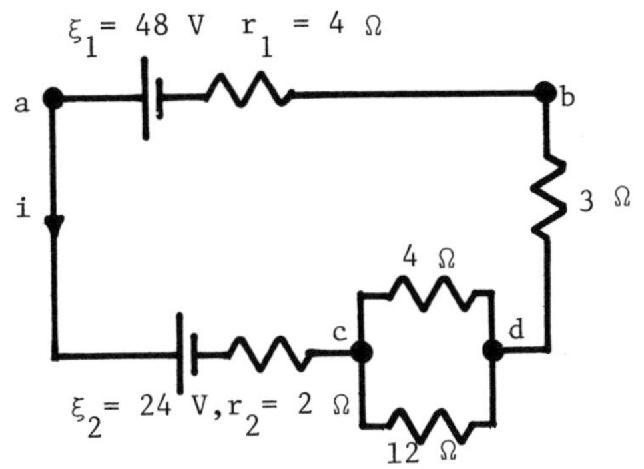

Fig. 3

6. In the circuit of Fig. 3, i is

(a) 0.96 A

(b) 2 A

(c) 6 A

7. In the circuit of Fig. 3, V_{ac} is

(a) 20 V

(b) 24 V

(c) 28 V

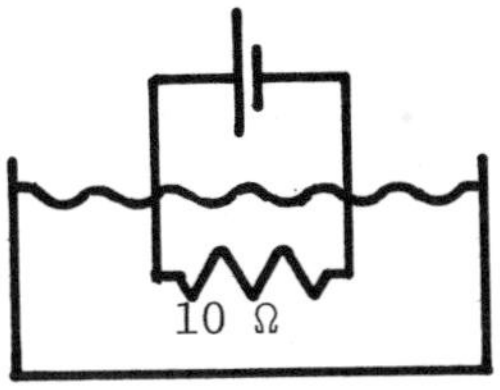

Fig. 4

8. A 10 ohm resistor is immersed in a container of 1 kg of water (Fig. 4). If it takes 4.18×10^3 J of energy to raise the temperature of 1 kg of water 1° C, to raise the temperature of the water 20°, you will have to run the heater

(a) 16.7 s

(b) 334 s

(c) 167.2 s

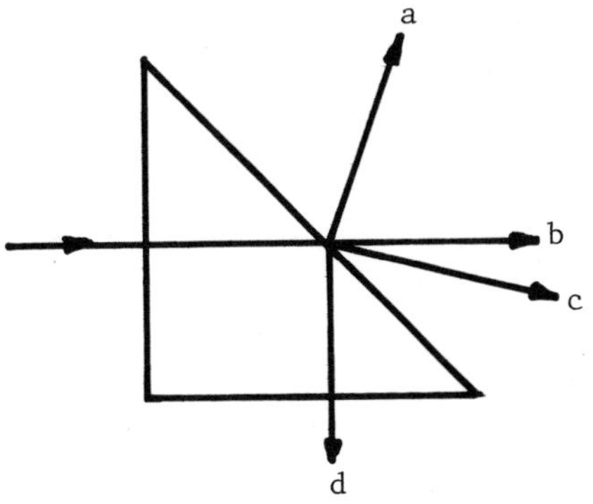

Fig. 5

9. A light ray enters a 45° prism, as shown in Fig. 5. If the ray emerges as ray d, the minimum index of refraction is

(a) 1.0

(b) 1.41

(c) 2.0

10. Imagine the ray emerges from the prism in Fig. 5. If it does, it emerges as

(a) Ray a

(b) Ray b

(c) Ray c

11. When an object is placed 12 cm to the left of a lens, a virtual image is formed 6 cm from the lens, on the same side of the lens as the object. The focal length of the lens is

(a) 4 cm

(b) -12 cm

(c) 12 cm

12. An object is placed 100 cm to the left of a converging lens of focal length f = 20 cm. A diverging lens of focal length f' = -10 cm is 20 cm to the right of the converging lens. The final image is formed

(a) 3.3 cm to left of diverging lens

(b) 10 cm to left of diverging lens

(c) 10 cm to right of diverging lens

CHAPTER 9. Transfer of Energy by Waves

Objectives

You should learn

1. the definition of the descriptive terms for wave motion, amplitude A, wavelength λ, period T, frequency f, and velocity v.

2. the difference between a plot of the displacement of points along a rope as a function of the distance along the rope and a plot of the displacement of one point on the rope as a function of time

3. the difference between the properties of a traveling wave and a standing wave

4. how to use the equation for a traveling wave

$$y = A \sin(2\pi s/\lambda \pm 2\pi t/T)$$

5. the equation for a standing wave

$$y = A(\sin 2\pi s/\lambda)\cos(2\pi t/T)$$

where A is the amplitude of the standing wave

6. the conditions for constructive and destructive interference

7. the meaning of a wave front

8. that both a double crest and a double trough result in a maximum disturbance

9. the meaning of phase difference

10. the condition for a minimum in the interference pattern for N sources

11. how to find a minimum in a diffraction pattern

Trouble Spots

The frequency f of a wave and the velocity v are not the same. The frequency is the number of vibrations a point in the medium makes in one second. The velocity of the wave is the distance moved by the wave in one complete vibration of the source. The frequency of the wave depends on the frequency of the source. The velocity of the wave depends on the medium. For example, the velocity of a wave along a rope

$$v = \sqrt{F/\mu}$$

where F is the tension in the rope and μ is the mass per unit length of the rope.

The relation between wavelength λ, frequency f, and velocity v is

$$\lambda \times f = v$$

The wavelength and frequency are inversely proportional. An increase in frequency corresponds to a decrease in wavelength and a decrease in frequency corresponds to an increase in wavelength.

When the source, that produces a traveling wave along a rope, vibrates with simple harmonic motion, each point on the rope vibrates with simple harmonic motion. Points along the rope have the same frequency and amplitude, but differ in phase. Points that are $\lambda/2$ apart vibrate 180° out of phase, points that are λ apart, vibrate 360° out of phase or, in other words, in phase.

For a standing wave, all points on a rope, except nodal points, vibrate with simple harmonic motion, but the points have different amplitudes. Maximum amplitude occurs at antinodal points. Nodal points are $\lambda/2$ apart. Points between nodal points vibrate in phase.

If you are given the number of the nodal line and the path difference, you can find the wavelength of the wave from the condition for a minimum

$$(n - 1/2)\,\lambda = \text{Path difference}$$

If you are given the number of the nodal line, the distance d between the two sources, and the angle θ_n between the nodal line and the center line, you can find the wavelength of the wave from the other condition for a minimum

$$(n - 1/2)\lambda = d \sin \theta_n$$

For N sources each separated by a distance d, the condition for a minimum is

$$m\,\lambda/Nd = \sin \theta_m$$

where m is an integer that is not a multiple of N. For example for three sources and m = 3

$$3\,\lambda/3d = \sin \theta_1$$

$$\lambda/d = \sin \theta_1$$

is the condition for the first maximum, n = 1, not a minimum.

The condition for a maximum for an interference pattern between sources a distance d apart is

$$n\,\lambda = d \sin \theta_n$$

The condition for a minimum for a diffraction pattern of slit width w is

$$m\,\lambda = w \sin \theta_m$$

The integral number of wavelengths on the left side of the above equation does not magically imply that there is a path difference of an integral number of wavelengths for destructive interference for a diffraction pattern.

You can see this by referring to Fig. 9.1. We divide the slit width w into eight sources.

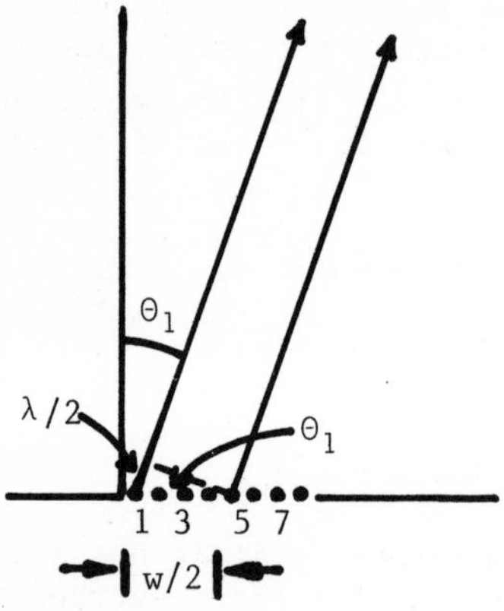

Fig. 9.1

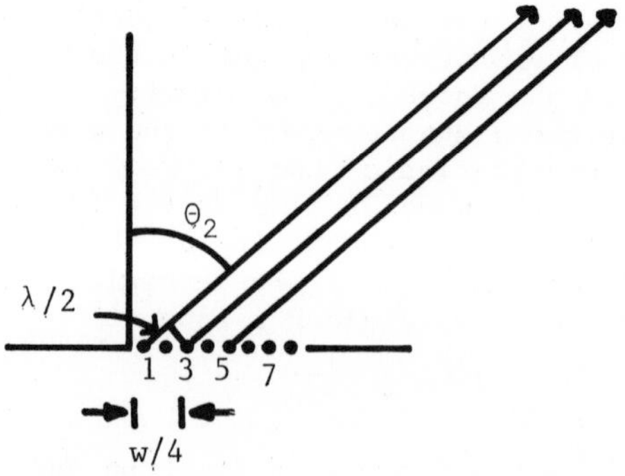

Fig. 9.2

We choose angle θ_1, for the first minimum, such that the path difference between source 1 and source 5 is $\lambda/2$. For a screen far from the slit, the waves from these sources arrive 180° out of phase and cancel. Similarly, source 2 cancels source 6, source 3 cancels source 7, and source 4 cancels source 8.

In Fig. 9.1

$$\frac{\lambda/2}{w/2} = \sin\theta_1$$

Because only one-half the slit width forms the hypotenuse of the triangle, the two's in the above equation cancel, and the condition for the first minimum is

$$\lambda = w \sin\theta_1$$

By dividing the slit into four parts such that there is a path difference of $\lambda/2$ between 1 and 3, between 2 and 4, between 5 and 7, between 6 and 8, we see from Fig. 9.2 that

$$\frac{\lambda/2}{w/4} = \sin\theta_2$$

or

$$2\lambda = w \sin\theta_2$$

A path difference of $\lambda/2$ results in destructive interference for both interference and diffraction patterns.

Diffraction is really the result of interference of a great number of sources. For historical reasons, we talk about interference when we deal with a finite number of sources. We use the term diffraction to describe interference between a very large number of sources.

Illustrative Examples

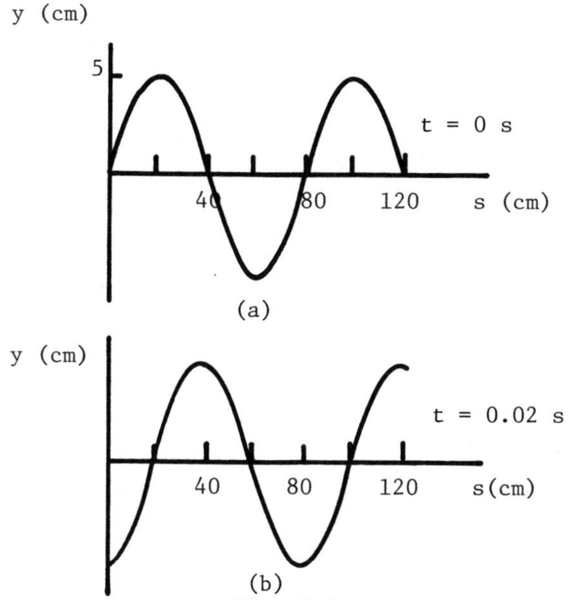

Fig. 9.3

1. Figure 9.3(a) is a plot of the displacement y as a function of the distance s from the source at t = 0. At t = 0.02 s, the rope, for the first time, looks like Fig. 9.3(b). Find (a) the amplitude, wavelength, period, frequency, and velocity of the wave. (b) Write the equation of the wave. Find (c) the displacement of the rope at s = 80 cm at t = 0.06 s and (d) the acceleration of the rope at s = 20 cm at t = 0.

   (a) From Fig. 9.3(a), we see that the amplitude A of the wave is 5 cm. In Fig. 9.3(a), the crests are 80 cm apart, thus λ= 80 cm. In Fig. 9.3(b), the crest has traveled 20 cm or a distance of $\lambda/4$. The wave travels $\lambda/4$ in a time T/4. So

$$T/4 = 0.02 \text{ s}$$

$$T = 0.08 \text{ s}$$

The frequency

$$f = 1/T$$

$$= 1/0.08 \text{ s}$$

$$= 12.5 \text{ s}^{-1}$$

The velocity

$$v = \lambda f$$

$$= (80 \text{ cm})(12.5 \text{ s}^{-1})$$

$$= 1000 \text{ cm s}^{-1}$$

(b) For a wave traveling to the right,

$$y = A \sin (2\pi s/\lambda - 2\pi t/T)$$

For this case with A = 5 cm, λ = 80 cm and T = 0.08 s,

$$y = 5 \text{ cm} \sin(\pi s/40 - 25\pi t)$$

(c) t = 0.06 s = (3/4)T. In 3/4 T, the crest moves

$$3/4\ \lambda = 3/4 \times 80 \text{ cm}$$

$$= 60 \text{ cm}$$

Since the crest is at s = 20 cm at t = 0, at t = 0.06 s it will be at s = 80 cm. Thus the displacement is y = 5 cm.

Alternatively, for s = 80 cm and t = 0.06 s,

$$y = 5 \text{ cm} \sin(\pi\ 80/40 - 25\pi \times 0.06)$$

$$= 5 \text{ cm} \sin(2\pi - 1.5\pi)$$

$$= 5 \text{ cm} \sin \pi/2$$

$$= 5 \text{ cm} \sin 90^o$$

$$= 5 \text{ cm} \times 1$$

$$= 5 \text{ cm}$$

(d) $a/y = -4\pi^2/T^2$

$$a = -4\pi^2 y/T^2$$

$$= -4\pi^2(5\ \text{cm})/(0.08\ \text{s})^2$$

$$= -3.12 \times 10^3\ \pi^2\ \text{cm s}^{-2}$$

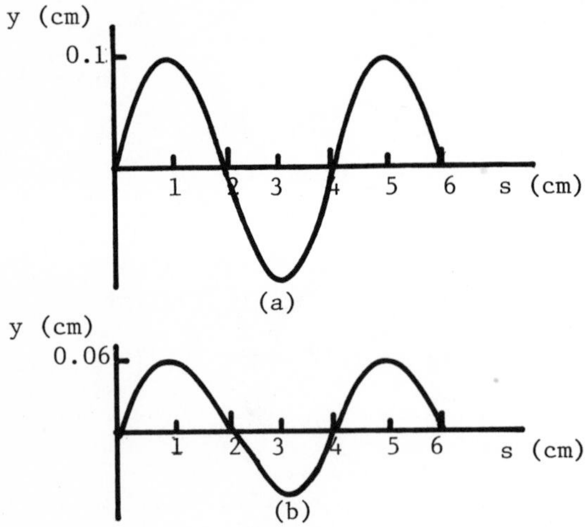

Fig. 9.4

2. The displacement of a rope looks like Fig. 9.4(a) at t = 0. At time t = 0.148 s for the first time, it looks like Fig. 9.4(b). (a) Is the wave along the rope a traveling wave or a standing wave? (b) What is the period of the wave? (c) Write the equation of the wave.

(a) Since points on the rope at s = 0, s = 2 m, and s = 4 m, remain at rest, this must be a standing wave.

(b) Since the point at s = 1 m has maximum displacement at t = 0, the equation of motion for this point is that for simple harmonic motion, that is,

$$y = 0.1\ \text{m} \cos 2\pi t/T$$

at t = 0.148 s

$$0.06\ \text{m} = 0.1\ \text{m} \cos 2\pi t/T$$

$$0.6 = \cos 2\pi t/T$$

Since $\cos 53^\circ = 0.6$ and $53^\circ = 0.93$ radians,

$$\cos 0.93 = \cos 2\pi t/T$$

or $\quad 0.93 = 2\pi t/T$

and $\quad T = 2\pi t/0.93$

For t = 0.148 s,

$$T = 2\pi(0.148\ \text{s})/0.93$$

$$= 0.10\ \text{s}$$

(c) The distance between nodes is $\lambda/2 = 2$ m, so $\lambda = 4$m. The amplitude of the standing wave is 0.1 m and its period T = 0.10 s. In general for a standing wave,

$$y = A(\sin 2\pi s/\lambda)\cos 2\pi t/T$$

For this case,

$$y = 0.1\ \text{m}(\sin \pi s/2) \cos 20\pi t$$

3. Two sources in a ripple tank separated, by a distance d, send out waves of wavelength λ. Find the angle for the nth maximum when (a) the sources have a phase difference of zero (b) the sources have a phase difference of 180°.

(a) For a phase difference of zero, the condition for a maximum is

$$n\lambda = d \sin \theta_n$$

(b) For a phase difference of 180°, one source sends out a crest, when the other sends out a trough. The conditions for maxima and minima when the sources are 180° out of phase are reversed. For this case the condition for a maximum is

$$(n - 1/2\ \lambda = d \sin \theta_n$$

4. Light of wavelength $\lambda$is incident upon two slits separated by a distance d. On a screen a long distance L away, an interference pattern is found. What is the distance between the maxima on the screen?

The condition for a maximum is

$$n\lambda = dD_n/L \qquad (1)$$

where D_n is the distance from the center of the screen to the nth maximum. For the (n + 1) maximum, the condition is

$$(n + 1)\lambda = dD_{n+1}/L \qquad (2)$$

The distance between adjacent maxima is $(D_{n+1} - D_n)$. From Equation 2,

$$D_{n+1} = (n + 1)\lambda L/d$$

From Equation 1,

$$D_n = n\lambda L/d$$

So,

$$D_{n+1} - D_n = (n + 1)\lambda L/d - n\lambda L/d$$

$$= \lambda L/d \quad (n + 1 - n)$$

$$= \lambda L/d$$

The condition for a minimum is

$$(n - 1/2)\lambda = dD_n'/L$$

where D_n' is the distance from the center of the screen to the nth minimum. You should be able to show that the separation between minima is also $\lambda L/d$.

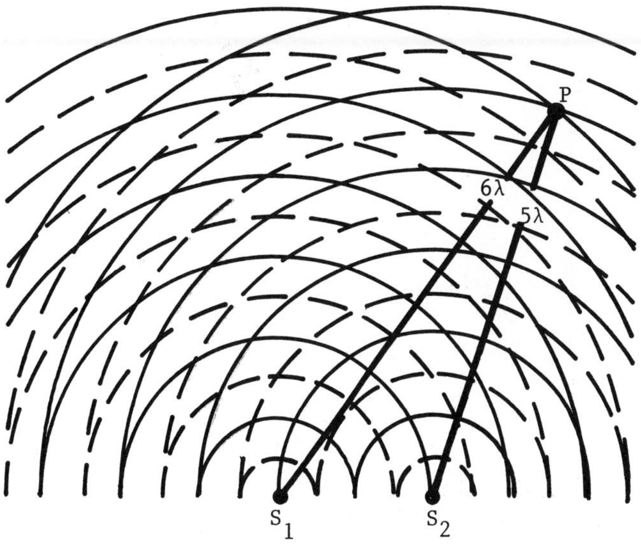

Fig. 9.5

5. In Fig. 9.5, both sources are in phase and send out waves of wavelength λ. The solid circles represent wave fronts through crests; the dashed circles represent wave fronts through troughs. At the moment shown, both S_1 and S_2 are sending out crests. What is (a) the distance from S_1 to P in terms of λ, (b) the distance from S_2 to P in terms of λ. (c) Is point P on a nodal or antinodal line?

(a) Counting the number of crests from S_1 to P, you find

$$S_1P = 6\lambda$$

(b) Counting the number of crests from S_2 to P, you find

$$S_2P = 5\lambda$$

(c) $S_1P - S_2P$ = Path difference

$6\lambda - 5\lambda$ = Path difference

Since the path difference is an integral

number of wavelengths, P is on an antinodal line.

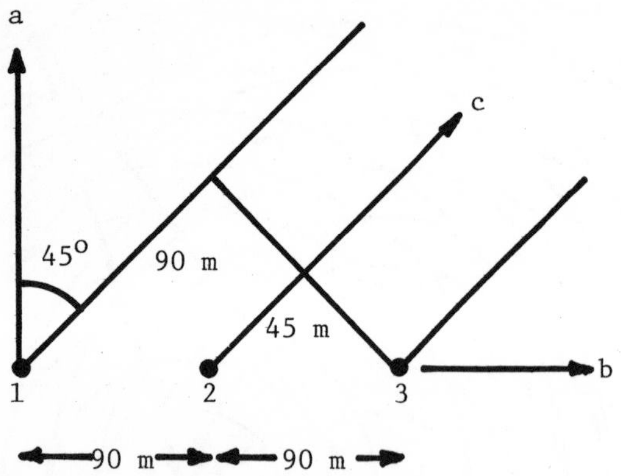

Fig. 9.6(a)

6. Three antennas emit waves of wavelength $\lambda = 180$ m of equal amplitude A_o and in phase. Compare the intensity of the resultant waves a long distance from the antennas in the directions a, b, and c in Fig. 9.6(a).

For a point far from the antennas, the distance from each antenna to that point, in the direction of a, is the same. For $A_o = A_1 = A_2 = A_3$, the total amplitude, as shown in Fig. 9.6(b) is $3\ A_o$.

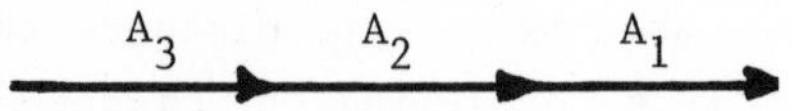

Fig. 9.6(b)

In the direction of b, waves from antenna 2 must travel 90 m = $\lambda/2$ farther than those from antenna 3 and waves from antenna 1 must travel 180 m = λfarther than those from antenna 3. In other words, waves from antenna 2 arrive 180° out of phase with those from antenna 3 and those from antenna 1 arrive 360° out of phase with those from antenna 3. Of course, a phase difference of 360° is equivalent to a phase difference of 0°. The resultant amplitude, as is shown in Fig. 9.6(c), is A_o.

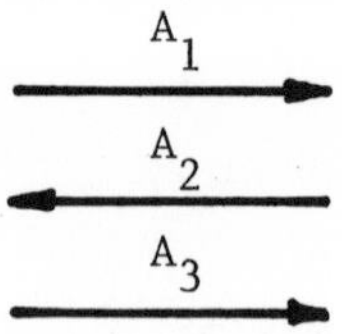

(c)

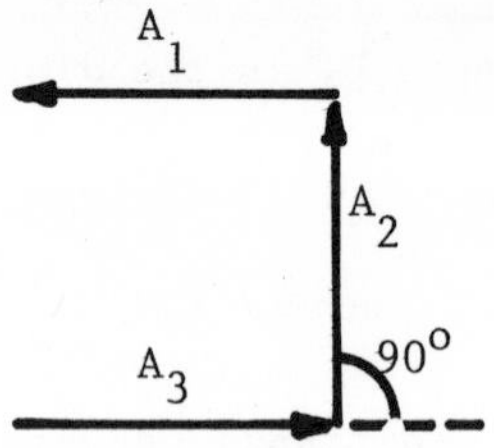

(d)

In direction c, the path difference between antenna 2 and antenna 3 is 45 m or $\lambda/4$ and the phase difference is 90°. A corresponding phase difference occurs between the waves from antenna 1 and antenna 2. As shown in Fig. 9.6(d) the resultant amplitude is $A_2 = A_o$.

Since intensity is proportional to the square of the amplitude, the intensity in direction a is proportional to $9A_o^2$, in direction b to A_o^2, and in direction c to A_o^2.

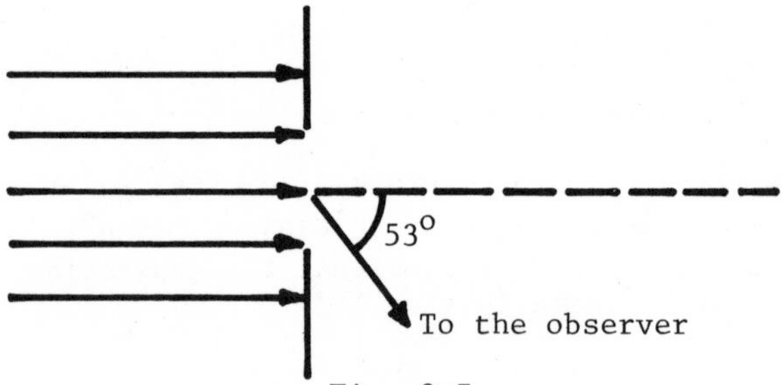

Fig. 9.7

7. Light of wavelength $\lambda = 5.00 \times 10^{-7}$ m is incident upon a slit of width w = 12.5 x 10^{-7} m. An observer far from the slit observes the light at an angle of 53°, as shown in Fig. 9.7. (a) Find the intensity of the light viewed by the observer. (b) The size of the slit is reduced to 10×10^{-7} m. Does the observer see a greater or less intensity of light?

(a) The condition for a minimum is

$$m\ \lambda = w \sin \theta_m$$

or

$$m = w \sin \theta_m / \lambda$$

For $w = 12.5 \times 10^{-7}$ m, $\lambda = 5.00 \times 10^{-7}$ m and $\theta = 53^{\circ}$,

$$m = \frac{12.5 \times 10^{-7} \text{ m} \times 4/5}{5.00 \times 10^{-7} \text{ m}}$$

$$= 2$$

This corresponds to a position of a minimum.

(b) Now for $w = 10^{-7}$ m

$$m = \frac{10^{-7} \text{ m} \times 4/5}{5.00 \times 10^{-7} \text{ m}}$$

$$= 1.6$$

Since m is <u>not</u> an integer, the observer no longer sites along a minimum and the intensity of the light is increased.

Problems

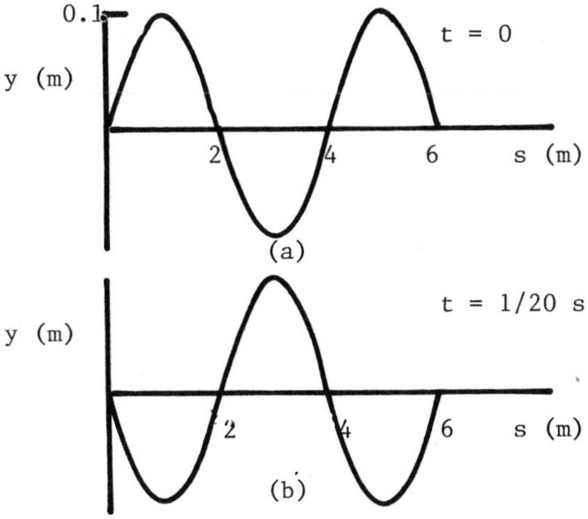

Fig. 9.8

1. Figure 9.8(a) is a plot of displacement as a function of distance s along a rope from the source of the waves at t = 0. For the first time after t = 0, the wave looks like Fig. 9.7(b) at t = 1/20 s. The period of the wave is

   (a) 1/40 s

   (b) 1/20 s

   (c) 1/10 s

2. In Problem 1, the velocity of the wave is

   (a) 0.4 m s^{-1}

   (b) 40 m s^{-1}

   (c) 80 m s^{-1}

3. In Problem 1, the equation of the wave is

   (a) $y = 0.1 \text{ m} \sin(\pi s/2 + 20\pi t)$

   (b) $y = 0.1 \text{ m} \sin(\pi s - 10\pi t)$

   (c) $y = 0.1 \text{ m} \sin(\pi s/2 - 20\pi t)$

   (d) $y = 0.1 \text{ m} (\sin \pi s/2) \cos 20\pi t$

4. In Fig. 9.9 sketch the displacement y as a function of time t for the piece of the rope at s = 1 m for Problem 1.

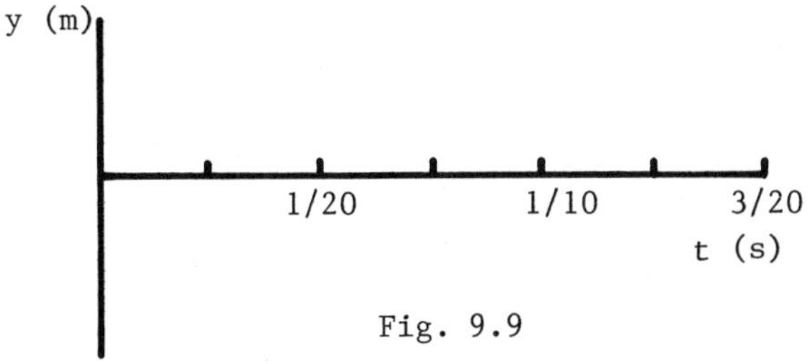

Fig. 9.9

5. Imagine that the rope of Problem 1 looked for the first time like Fig. 9.10 at time t = 1/20 s.

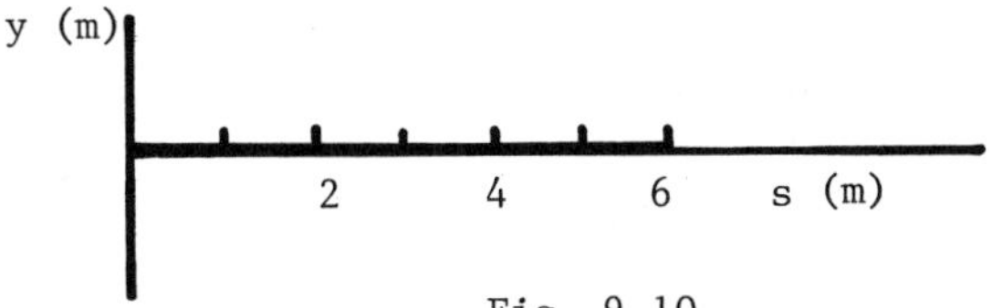

Fig. 9.10

The equation of the wave is

(a) $y = 0.1$ m $\sin(\pi s/2 - 20\pi t)$

(b) $y = 0$

(c) $y = 0.1$ m$(\sin \pi s/2) \cos 20\pi t$

(d) $y = 0.1$ m$(\sin \pi s/2) \cos 10\pi t$

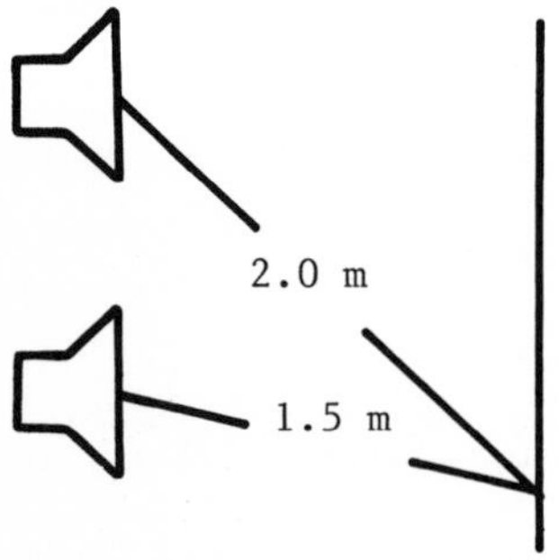

Fig. 9.11

6. A person standing on a line, that is the perpendicular bisector of the line between two loud speakers, hears maximum intensity of sound. When he moves parallel to the line between the speakers and he is 1.5 m from one speaker and 2.0 m from the other (Fig. 9.11), he reaches for the first time a point of minimum intensity of sound. The velocity of sound is 330 m s^{-1}. The frequency emitted by the speakers is

(a) 220 s^{-1}

(b) 330 s^{-1}

(c) 660 s^{-1}

7. In a double slit experiment the distance between adjacent bright lines is 0.01 cm. If the distance to the screen is doubled, the distance between the slits is halved, and the wavelength of the light is changed from 4.0×10^{-5} cm to 6.5×10^{-5} cm, the new distance between the bright lines is

(a) 0.01 cm

(b) 0.01625 cm

(c) 0.065 cm

(d) 0.130 cm

8. Light of wavelength $\lambda = 7 \times 10^{-7}$ m is incident upon 4 slits. The distance between each slit is 12.5×10^{-7} m. The light is viewed at an angle of 16° from the central line. In this direction, you see

(a) a principal maximum

(b) a secondary maximum

(c) a minimum

9. Light of wavelength $\lambda$is incident upon a slit of width $w = 2.2 \times 10^{-6}$ m. On a screen 2 m from the slit, the width of the central maximum is 1 m. The wavelength λ is

(a) 2.75×10^{-7} m

(b) 5.5×10^{-7} m

(c) 11.0×10^{-7} m

10. Light with wavelength $\lambda = 5.0 \times 10^{-5}$ cm passes through a slit and forms a diffraction pattern on a screen 100 cm away. The slit width for which the ratio of the central maximum to slit width w is 2 is

(a) 5×10^{-2}

(b) 7.1×10^{-2}

(c) 10×10^{-2}

11. You wish to use a microscope that will distinguish between objects that are very close together. If you have your choice of light to use, you would use

(a) radio waves

(b) infrared light

(c) visible light

(d) ultraviolet light

CHAPTER 10. Electromagnetic Induction and Electromagnetic Waves

Objectives

You should learn

1. that a changing magnetic field produces an induced current

2. that the direction of the induced current is such to oppose the change that produced it

3. the definition of magnetic flux ϕ_B

4. how to show that $\Delta\phi_B / \Delta t = \xi = B\ell v$ for a wire of length ℓ moving with velocity v in a magnetic field $\vec{B}$, perpendicular to $\vec{v}$

5. that a changing magnetic field produces an electric field

6. that a changing electric field produces a magnetic field

7. that an electromagnetic wave carries energy that is proportional to E^2 or B^2

8. that the momentum of an electromagnetic equals the energy of the wave divided by the speed of light, that is,

$$p = \frac{\text{Energy}}{c}$$

Trouble Spots

The direction of an induced current, found by experiment, is a consequence of conservation of energy. For example, once the direction of deflection of a galvanometer is checked with a known source, we know that a deflection toward the left indicates that positive charge must flow in the coil as shown in Fig. 10.1, when the south-seeking pole is moved away from the coil.

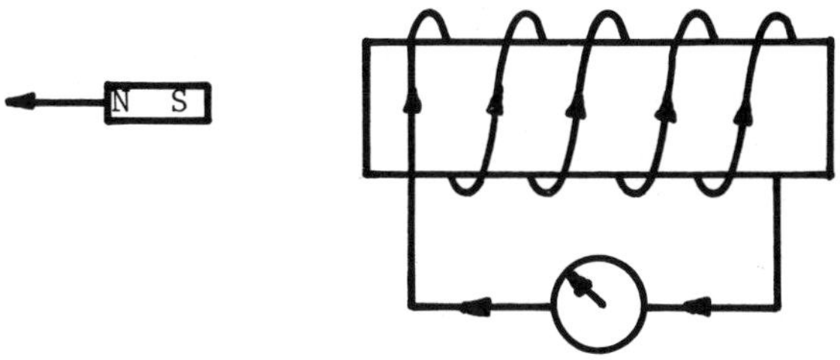

Fig. 10.1

The current in the face of the coil closest to the magnet is in a counterclockwise direction. Thus, this face acts like a north-seeking pole. This is a result of conservation of energy. When you remove the south-seeking pole from the coil, you induce a current in the coil. To conserve energy you must do work. You will do work if the face of the coil closest to the south-seeking pole of the magnet acts like a north-seeking pole. You must exert a force through a distance to remove a south-seeking pole from the face of a coil that acts like a north-seeking pole.

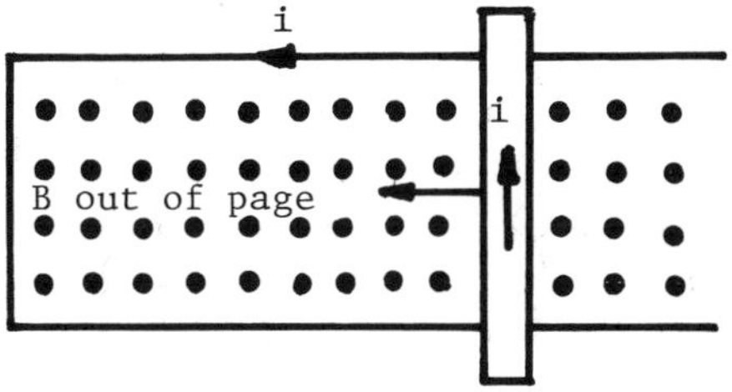

Fig. 10.2

There are a number of ways to find the direction of the current in the loop of Fig. 10.2 when the bar is moved to the left. For example, when the bar is moved to the left, the number of magnetic field lines through the loop decreases. To oppose the change, the induced current must be in such a direction that it produces a magnetic field inside the loop that is out of the page. This will be true for a counterclockwise induced current. Alternatively to produce the current, you must do work, that is, you must exert a force to move the bar to the left. Your force must oppose the force of the magnetic field. Thus, the force on the bar due to the magnetic field must be to the right. By the second right hand rule, you see that the magnetic field out of the page will exert a force on the bar to the right if the current in the bar is up.

The induced emf in a rotating coil is not a maximum when the number of field lines through the coil is a maximum. The emf equals the rate of change of magnetic flux $\Delta\phi_B$ not the magnetic flux. Just as the acceleration of a pendulum bob is a maximum when the velocity changes from 0 to some finite value, the emf is a maximum when the number of field lines changes from 0 to some finite value. See Fig. 10.3

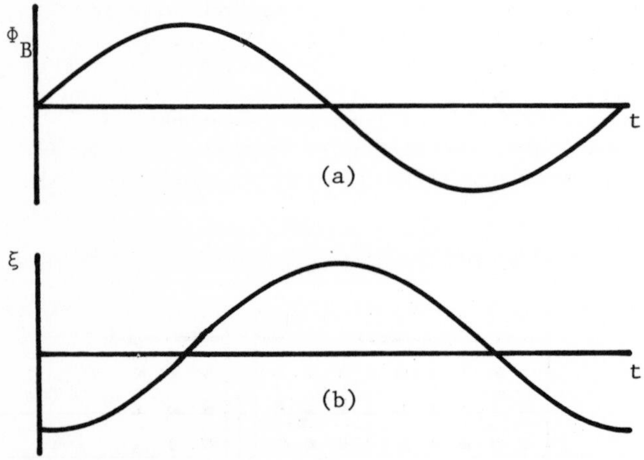

Fig. 10.3

While the electric and magnetic fields at a point near oscillating charges are not in phase, they are in phase at points far from the charges. The magnetic field is a maximum when the electric field is a maximum. The magnetic field is always perpendicular to the electric field.

Illustrative Examples

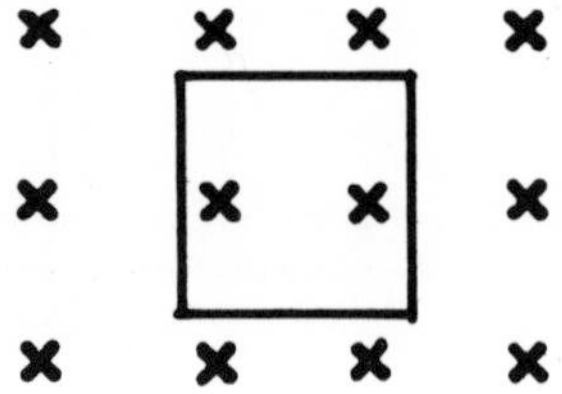

Fig. 10.4

1. A square loop of wire, 1 m on a side, has a resistance of 0.5 ohm. Initially a magnetic field of intensity B = 0.10 $N(A\text{-}m)^{-1}$ points into the page. If this field is reduced in magnitude to 0.05 $N(A\text{-}m)^{-1}$ at a constant rate, in 2 seconds, what electric current (Fig. 10.4) is produced in the loop? Give magnitude and direction of the current.

The induced emf equals the rate of change of the magnetic flux,

$$\xi = \Delta\phi_B/\Delta t$$

$$\xi = \Delta(BA)/\Delta t$$

$$\xi = A\Delta B/\Delta t$$

$$\xi = (1\ m^2)\ 0.5\ N(A\text{-}m)^{-1}/2\ s$$

$$\xi = 0.25\ V$$

In the above equation, we took A out from the delta symbol because only the magnetic field changes.

$$i = \xi/R$$

$$= 0.25\ V/0.5\ ohm$$

$$= 0.5\ A$$

To resist the decrease in the magnetic field inside the coil that is into the page, the induced current is in a clockwise direction.

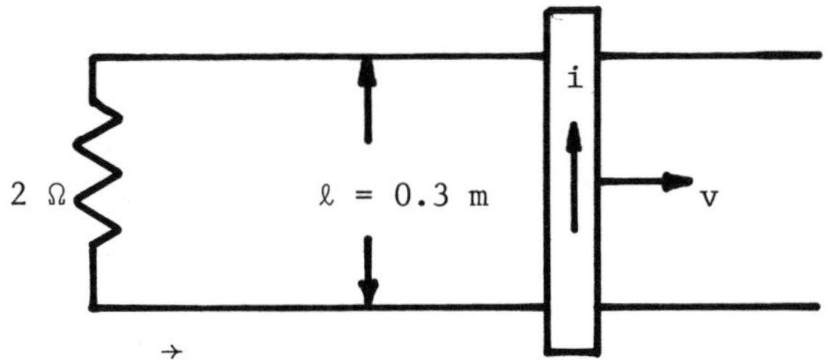

$\vec{B}$ everywhere into the page

Fig. 10.5

2. A metal rod rest on horizontal rails that are separated by a distance ℓ = 0.3 m and connected on the right by a resistor with resistance R = 2 ohm. A magnetic field B = 0.5 $N(A\text{-}m)^{-1}$ points vertically downward. If the rod is moved to the left with v = 4.0 m s^{-1}. Find (a) the induced emf, (b) the current through the rod, (c) the force on the rod (d) the mechanical power supplied to the rod, and (e) the electric power dissipated in R.

(a) $\xi = B\ell v$

$= 0.5\ N(A\text{-}m)^{-1} \times 0.3\ m \times 4.0\ m\ s^{-1}$

$= 0.6\ V$

(b) $i = \xi/R$

$= 0.6\ V/2\ ohm$

$= 0.3\ A$

The direction of the current is as shown in Fig. 10.4. When the rod is moved to the right, there is an increase in the magnetic field into the page through the loop. If the current is in a counter-clockwise direction, it produces a magnetic field inside the loop that is out of the page.

(c) $F_m = Bi\ell$

$= 0.5\ N(A\text{-}m)^{-1} \times 0.3\ A \times 0.3\ m$

$= 0.045\ N$

Using the second right hand rule, you see that when the current is up, the magnetic field into the page, the magnetic force is to the left.

(d) The mechanical work

$P_m = \Delta\ Work/\Delta\ t$

$= F\ \Delta\ s/\Delta t$

$= Fv\Delta t/\Delta t$

$= Fv$

$= 0.045\ N \times 4.0\ m\ s^{-1}$

$= 0.18\ W$

(e) The electric power dissipated as heat

$P = i^2R$

$= (0.09\ A^2)(2\ ohm)$

$= 0.18\ W$

Power, and therefore, energy = power x time, is conserved. The power produced by the mechanical work goes into the power dissipated as heat.

3. Suppose that a region of width 4 m has a magnetic field B of magnitude 0.5 $N(A\text{-}m)^{-1}$ pointing into the page, and that B = 0 elsewhere. If a square loop of wire 2 m on a side, is pulled through this region at a constant velocity of 2 m s^{-1} to the right in Fig. 10.6(a). The loop has a resistance R = 0.1 ohm. Find the current in the wire as a function of time, letting Fig. 10.6(a) define positive current. At t = 0, one-half the loop is in the magnetic field.

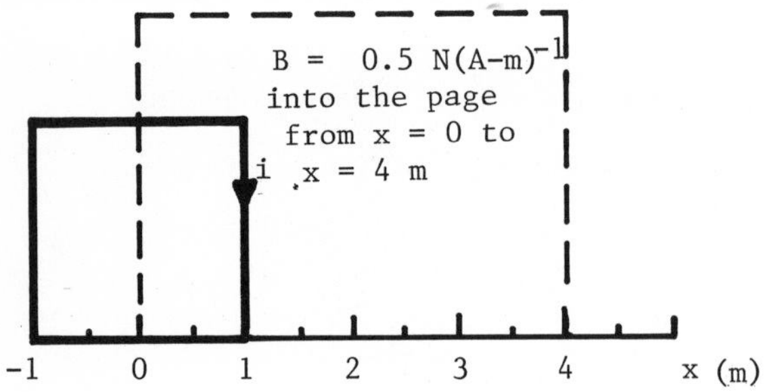

Fig. 10.6(a)

First we find the magnetic flux as a function of time. At t = -1 s, the center of the loop is at x = -2 m and no magnetic field lines pass through the coil. Thus at t = -1 s, $\phi_B = 0$. At t = -0.5 s, the center of the coil is at x = -1 m and again $\phi_B = 0$. At t = 0, the magnetic field lines thread half the coil and

$$\phi_B = B(A)$$

$$= 0.5 \text{ N(A-m)}^{-1} \times 2 \text{ m}^2$$

$$= 1.0 \text{ N-m A}^{-1}$$

At t = 0.5 s the center of the loop is at x = 1 m and the magnetic field threads the entire loop. Now

$$\phi_B = 0.5 \text{ N(A-m)}^{-1} \times 4 \text{ m}^2$$

$$= 2.0 \text{ N-m A}^{-1}$$

The entire loop is in the coil from t = 0.5 s to t = 1.5 s so during that time the flux remains constant at 2.0 N-m A^{-1}. At t = 2.0 s one-half the loop is in the field and ϕ_B = 1.0 N-m A^{-1}. At t = 2.5 s the loop has left the field and $\phi_B = 0$. The values of ϕ_B at different times is given in Table 10.1. A plot of ϕ_B as a function of time is given in Fig. 10.6(b).

Table 10.1

t (s)	ϕ_B (N-m A^{-1})
-1	0
-0.5	0
0	1
0.5	2
1.0	2
1.5	2
2.0	1
2.5	0
3.0	0

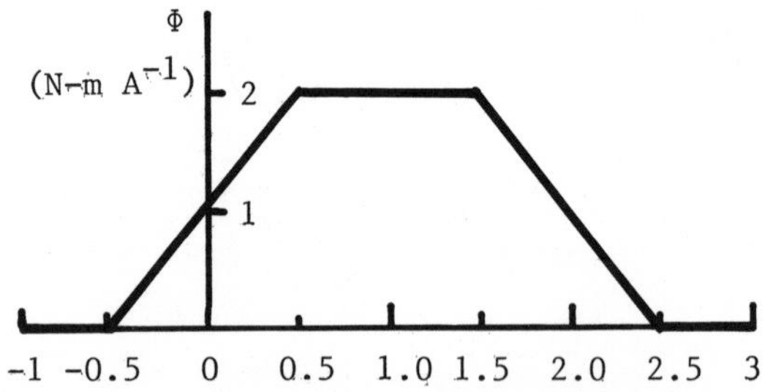

(b)

The induced emf is found from the slope of ϕ_B versus t. For example, from t = -1 s to t = -0.5 s, $\Delta\phi_B = 0$ and $\xi = 0$. From t = -0.5 s to t = 0 s, $\Delta\phi_B$ = 1 N-m A^{-1} and

$$\xi = \Delta\phi_B/\Delta t$$

$$= 1 \text{ N-m A}^{-1}/(0 - (-0.5)) \text{ s}$$

$$= 2 \text{ V}$$

Since the slope of ϕ_B versus t remains constant from t = -0.5 s to t = 0.5 s, the emf remains constant and equal to 2 V. The slope of ϕ_B versus t is zero from t = 0.5 s to t = 1.5 s. Thus the emf is zero from t = 0.5 s to t = 1.5 s. The magnitude of the slope (and of the emf) is 2.0 V from t = 1.5 s to t = 2.5 s. After t = 2.5 s, the slope again equals 0.

The induced current at any instant is $i = \xi/R$. For example, from t = 1.5 s to t = 2.5 s,

$$i = \xi/R$$

$$= 2 \text{ V}/0.1 \text{ ohm}$$

$$= 20 \text{ A}.$$

If i in Fig. 10.5(a) is positive, the current at t = 0 is negative. If the current were down the magnetic field would help, rather than hinder the motion of the loop to the right. For example, with a current down, and the magnetic field into the page, the magnetic force, by the second right hand rule, is to the right. Since we are moving the loop to the right, in order to do work, the magnetic field must exert a force on the loop to the left.

A plot of the current i as a function of time is given in Fig. 10.6(c).

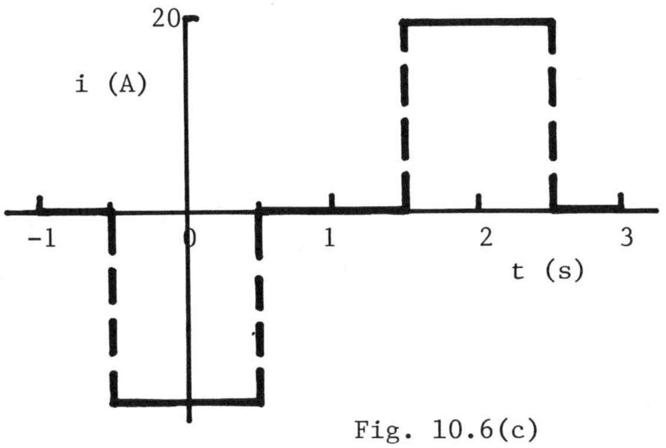

Fig. 10.6(c)

4. An interstellar spaceship of mass 4000 kg, starting at rest, emits 2×10^6 W of radiant power from its rear for one year (3×10^7 s). What velocity does it acquire as a result?

The energy of the electromagnetic waves equals the product of the power and the time, that is,

$$\text{Energy} = \text{Power} \times \text{time}$$

$$= 2 \times 10^6 \text{ W} \times 3 \times 10^7 \text{ s}$$

$$= 6 \times 10^{13} \text{ J}$$

The momentum of the electromagnetic waves emitted equals the energy of the waves divided by the speed of light:

$$p = \text{Energy}/c$$

$$= 6 \times 10^{13} \text{ J}/3 \times 10^8 \text{ m s}^{-1}$$

$$= 2 \times 10^5 \text{ kg-m s}^{-1}$$

To conserve momentum, the spaceship must acquire a momentum that is equal and opposite to the momentum of the electromagnetic waves. From conservation of momentum,

$$\text{Before } p = \text{After } p$$

$$0 = 2 \times 10^5 \text{ kg-m s}^{-1} + Mv$$

$$\text{or} \quad Mv = -\,2 \times 10^5 \text{ kg-m s}^{-1}$$

$$4000 \text{ kg } v = -\,2 \times 10^5 \text{ kg-m s}^{-1}$$

$$v = -\,50 \text{ m s}^{-1}$$

Problems

1. A north-seeking pole is moved toward a coil. The face of the coil toward which the north-seeking pole is moved acts like

   (a) a north-seeking pole

   (b) a south-seeking pole

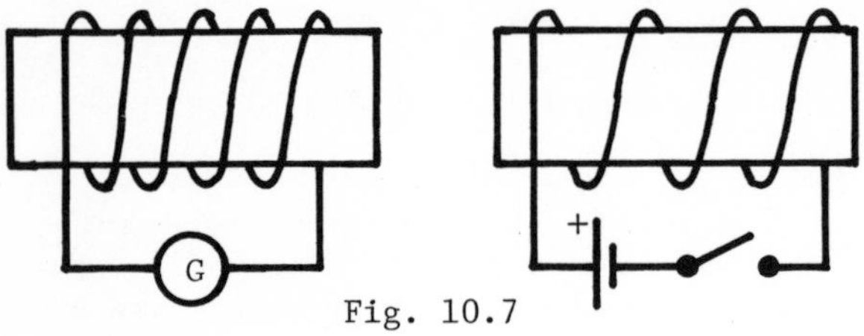

Fig. 10.7

2. In Fig. 10.7, when the switch in the primary is thrown, the face of the secondary closest to the primary acts like

   (a) a north-seeking pole

   (b) a south-seeking pole

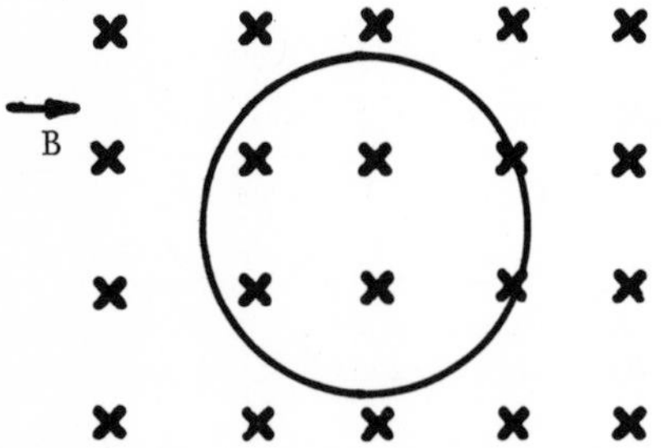

Fig. 10.8

3. A positive charge +q moves in a circle in a magnetic field that is into the page (Fig. 10.8). If the magnitude of the magnetic field is increased, the speed of the charge will

   (a) decrease

   (b) remain the same

   (c) increase

4. A rod rests on horizontal rails that are separated by a distance $\ell = 1$ m and connected on the left by a resistor with resistance R = 5 ohm.(Fig. 10.9). The rod is moved to the left with a velocity $v = 20\ \text{m s}^{-1}$. The magnitude of the magnetic field $B = 0.2\ \text{N(A-m)}^{-1}$.

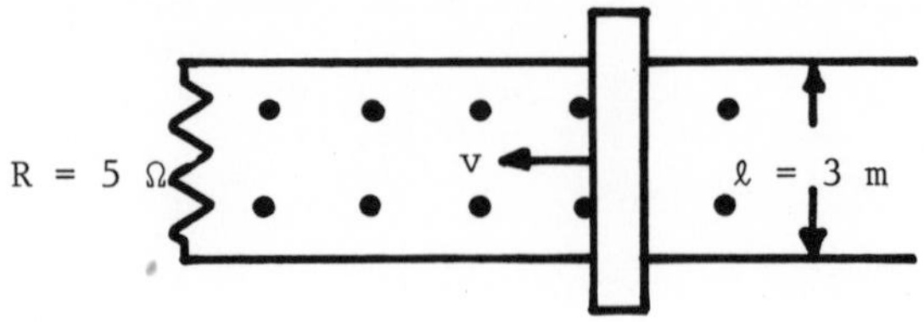

Fig. 10.9

The direction of the induced current in the loop is

   (a) clockwise

   (b) counterclockwise

5. In Problem 4, the magnitude of the current is

   (a) 0.8 A

   (b) 2.0 A

   (c) 4.0 A

6. In Problem 4, to move the rod with a constant velocity $v = 20\ \text{m s}^{-1}$ to the right, a person must exert a force of

   (a) 0.16 N to the left

   (b) 0.16 N to the right

   (c) 4.0 N to the left

   (d) 4.0 N to the right

7. A toy train transformer is plugged into a 120-V line. If the power supplied to the line is 120 watts, the current that exists in the 12-V secondary is

   (a) 1/10 A

   (b) 5 A

   (c) 10 A

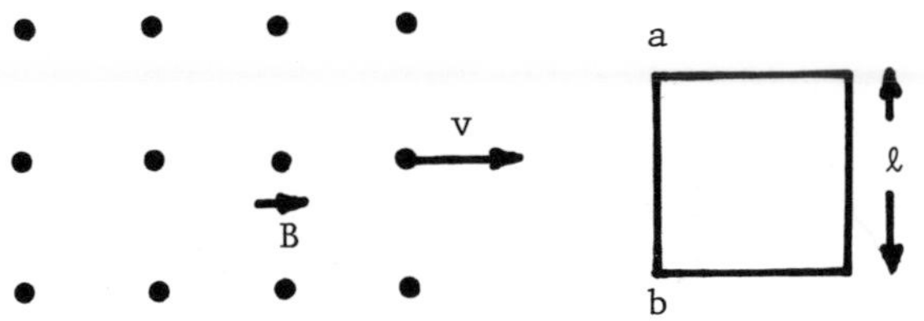

Fig. 10.10

8. A magnetic field, out of the page, moves to the right with velocity v toward a loop of area A and length ℓ (Fig. 10.10). This moving magnetic field B produces an electric field equal to

(a) Bℓv from a to b

(b) Bℓv from b to a

(c) Bv from a to b

(d) Bv from b to a

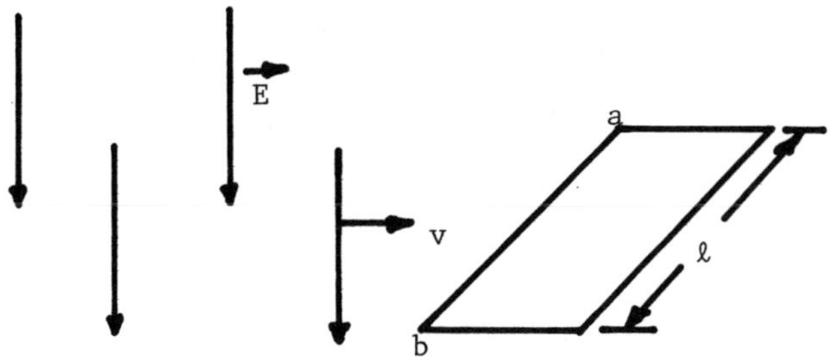

Fig. 10.11

9. An electric field E, vertically downward, moves to the right with velocity v towards a loop of area A and length ℓ(Fig. 10.11). This moving electric field E produces a magnetic field equal to

(a) B/v from a to b

(b) B/v from b to a

(c) Bc^2/v from a to b

(d) Bc^2/v from b to a

10. A certain electromagnetic wave delivers an energy of 3.3×10^{-20} J to a charged particle. The momentum associated with the wave is

(a) 1.1×10^{-28} kg m s^{-1}

(b) 3.3×10^{-20} kg m s^{-1}

(c) 9.9×10^{-12} kg m s^{-1}

Sample Quiz for Chapters 9 and 10

1. A source of frequency f produces a wave of velocity v in a certain medium. If the medium is changed so that the velocity is 2v

   (a) the frequency of the wave is doubled

   (b) the wavelength of the wave is halved

   (c) the wavelength of the wave is doubled

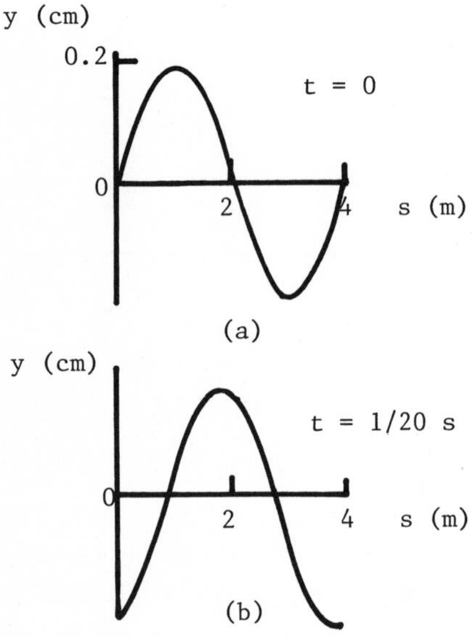

Fig. 1

2. A snapshot of a wave on a rope at t = 0 s looks like Fig. 1(a). At t = 1/20 s, for the first time it looks like Fig. 1(b). The velocity of the wave is

   (a) 0.8 m s^{-1}

   (b) 20 m s^{-1}

   (c) 40 m s^{-1}

3. The acceleration of the piece of rope at s = 1 m in Fig. 1(a) is

   (a) 0

   (b) $20\pi^2 \text{ m s}^{-2}$

   (c) $100\pi^2 \text{ m s}^{-2}$

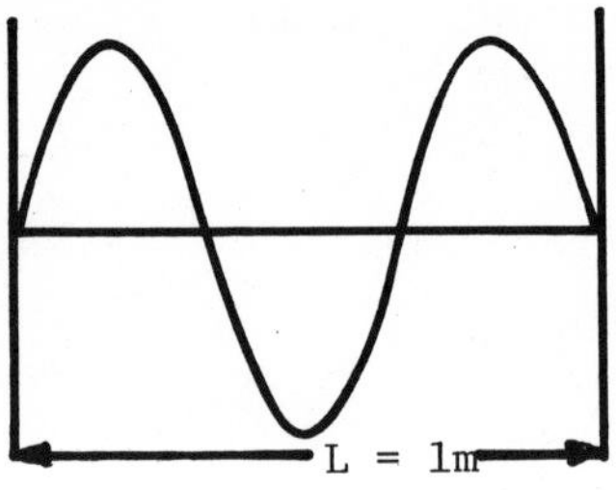

Fig. 2

4. A standing wave is set up in a rope that is fixed at both ends. At one instant of time, the rope looks like Fig. 2. The velocity of the wave is 100 m s^{-1}. The frequency of the wave is

   (a) 50 s^{-1}

   (b) 100 s^{-1}

   (c) 150 s^{-1}

   (d) 200 s^{-1}

5. Four sources are each separated by a distance d. When light of wavelength λ is incident upon them, the first maximum to the right of the center line is found at angle θ such that the sine of the angle θ equals

   (a) $\lambda/4d$

   (b) $\lambda/2d$

   (c) λ/d

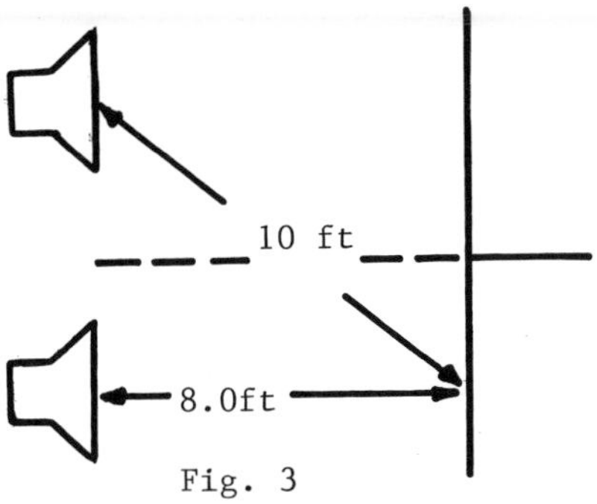

Fig. 3

6. A person is at equal distances from two speakers of a stereo hi-fi system and hears a note of a single frequency. He moves sideways until he hears the note fade to a minimum (Fig. 3). At this position, he is 10 ft from the left speaker and 8.0 ft from the right speaker. The speed of sound is 1100 ft s^{-1}. The frequency of the note is

(a) 275 s^{-1}

(b) 550 s^{-1}

(c) 4400 s^{-1}

7. Light of wavelength $\lambda = 5 \times 10^{-5}$ cm falls on a slit of width $w = 5 \times 10^{-3}$ cm and arrives at a screen a distance $L = 100$ cm away. In Fig. 4, sketch the intensity pattern as a function of the distance s along the screen.

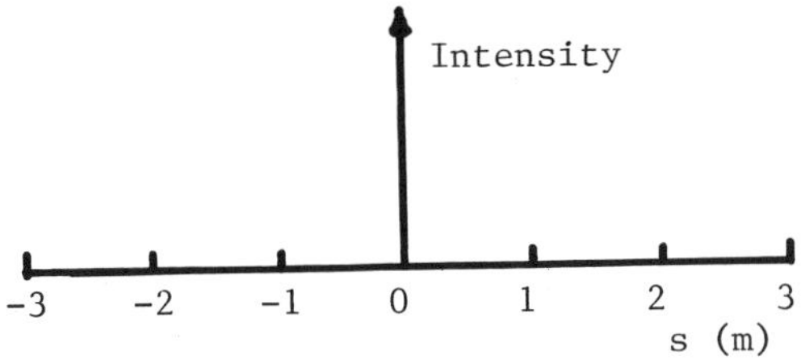

Fig. 4

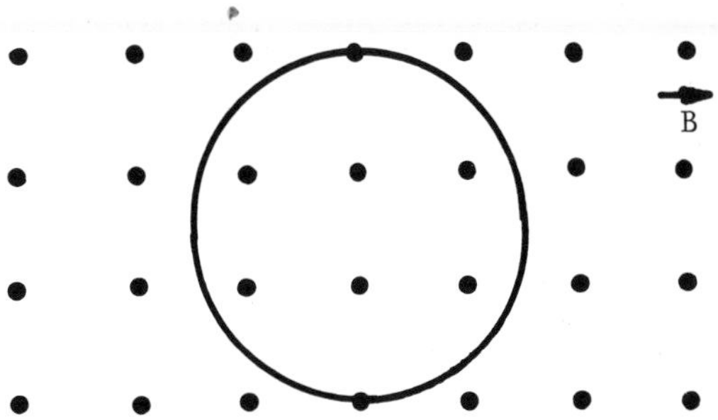

Fig. 5

8. A magnetic field $B = 0.2$ $N(A\text{-}m)^{-1}$ out of the page threads a coil of radius $1/\pi$ m and resistance $R = 0.1$ ohm (Fig. 5). The field is changed at a uniform rate so that in 0.2 s it has a magnitude of 0.4 $N(A\text{-}m)^{-1}$ into the page. The induced current is

(a) $10/\pi$ A

(b) $20/\pi$ A

(c) $30/\pi$ A

9. In Problem 8, the direction of the induced current is

(a) clockwise

(b) counterclockwise

10.

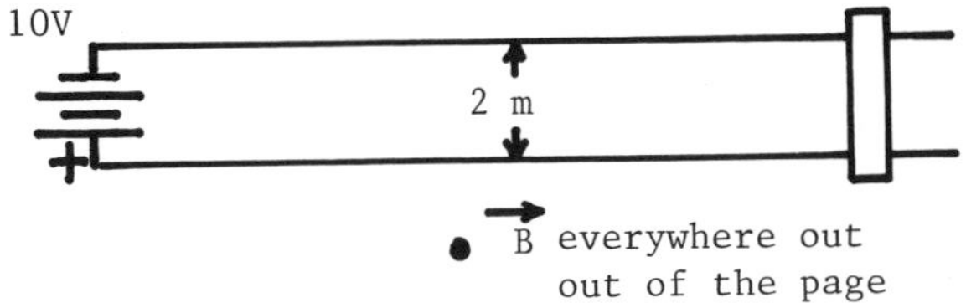

A conducting bar rests across railway tracks that are connected to each other only by a battery (Fig. 6). A uniform magnetic field $B = 0.1$ $N(A\text{-}m)^{-1}$ perpendicular to the track is everywhere. With the battery in place, B is increased from 0.1 $N(A\text{-}m)^{-1}$ to 1.0 $N(A\text{-}m)^{-1}$ at a constant rate over a period of 9 s. In order to experience no force while the field is changing, the bar must have been placed at rest a distance from the end of the rail of

(a) 50 m

(b) 100 m

(c) 200 m

(d) none of the above

CHAPTER 11. Relativity

Objectives

You should learn that

1. the velocity of light is the same for all observers

2. events, separated in space, that are simultaneous for one observer are not simultaneous for a second observer, who moves relative to the first observer

3. the shorter time interval between two events, the proper time, is measured by an observer for whom the two events happen in the same place

4. the shorter length of a rod, the proper length, is measured by an observer, who is at rest with respect to the rod

5. clocks synchronized by observers on a train are not synchronized to observers, who see the train in motion

6. the "additive" law for velocities does not hold at relativistic speeds

7. the space time interval

   $$(\Delta I)^2 = (\Delta t)^2 - (\Delta x/c)^2$$

   is the same for all observers

8. the relativistic expression for momentum is

   $$p = \frac{m_o u}{\sqrt{1 - u^2/c^2}}$$

   where u is the speed of the object and m_o its rest mass

9. the relativistic expression for the total energy of a particle is

   $$E = \frac{m_o c^2}{\sqrt{1 - u^2/c^2}}$$

10. the relativistic expression for kinetic energy is

$$KE = E - m_o c^2$$

11. energy and mass are interchangeable

12. in a decay process, you should include $m_o c^2$ as a part of the energy for an object that is at rest after the decay

Trouble Spots

The proper time interval between two events is read on a single clock and happens in the same place for that observer. For example, imagine that on July 4th as a spaceship is directly over Indianapolis at 12 noon, What will be the reading on the spaceship's clocks when they see the light from next year's fireworks? The observer in the spaceship records the time between the two events:

1. passing Indianapolis at 12 noon

2. the reception of the fireworks

on the same clock in the same place, but the observer in the spaceship does not record the time between the two events

1. passing Indianapolis at 12 noon

2. the sending off the fireworks

in the same place. The time interval between the latter two events is not a proper time interval for observers in the spaceship.

When measuring the length of an object by finding the time for an object to pass him, an observer measures a proper time, but an improper length. For example, in Fig. 11.1(a) an observer on a train records the time at which the left end of the rod in coordinate system S passes him. In Fig. 11.1(b) he records the time at which the right end of the rod passes him. Since the two ends of the rod pass him in the same place in his reference frame S', the time between the two events

1. the passing of the left end of the rod

2. the passing of the right end of the rod

is a proper time interval in reference frame S'. Since the rod moves relative to S', the length of the rod L is not a proper length in reference frame S'.

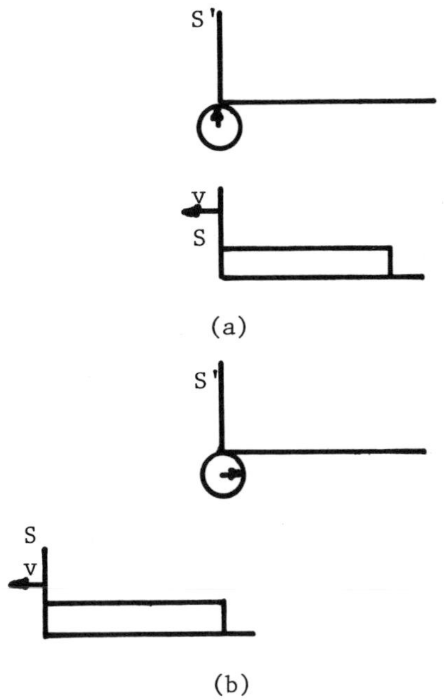

Fig. 11.1

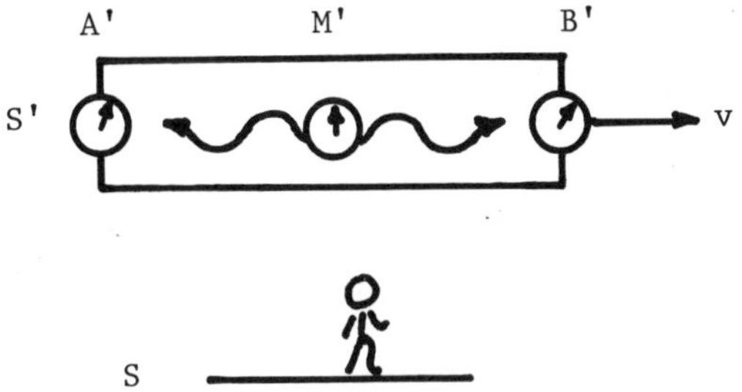

Fig. 11.2

Observers on a train synchronize their clocks by sending out a light signal from the master clock M' at t_0' (Fig. 11.2). Observers adjacent to clocks A' and B' set (but do not start) their clocks at $(t_0' + x'/c)$, where x' is the distance from the master clock M' to either A' or B', as measured by observers on the train. The clocks are started when they receive the light signal.

Observers on the platform do not agree that the clocks are synchronized. These observers say the rear clock moves toward the light signal and is started before $(t_0' + x'/c)$ while the front clock moves away from the light signal and is started after $(t_0' + x'/c)$ Observers on the platform say that the rear clock is set ahead and the front clock is set behind.

Measurements are reciprocal. For two trains S and S' moving relative to each other, observers in train S find that

1. the length of train S' is shorter than it is measured by observers on S'

2. clocks on S' run slow

3. the clock at the front of S' is set behind the clock at the rear of S'

Observers on train S' find that

1. the length of train S is shorter than it is measured by observers on S

2. clocks on S run slow

3. the clock at the front of S is set behind the clock at the rear of S

Imagine that you are told that trains S' and S" move relative to a platform with speeds u_1 and u_2 and are asked to find the speed of S" relative to S'. Let u_1 equal the relative speed v and use the velocity transformation to find the velocity of S" as measured by observers on S'.

When a constant force F acts on an object that is initially at rest,

$$Ft = \Delta p$$

$$= \frac{m_o u}{\sqrt{1 - u^2/c^2}}$$

where u is the speed of the object at time t. If the object moves a distance s in this time,

$$\text{Work done} = Fs$$

$$= \Delta E$$

$$= \frac{m_o c^2}{\sqrt{1 - u^2/c^2}} - m_o c^2$$

$$= m_o c^2 \left[\frac{1}{\sqrt{1 - u^2/c^2}} - 1 \right]$$

Never use classical expressions for momentum and kinetic energy of an object unless the speed of the object is much less than the speed of light. In applying conservation of energy for a decay process, don't forget to include the rest mass energy of one of the particles if it is at rest after the decay.

You will find the following relations useful:

$$E = \sqrt{(pc)^2 + (m_o c^2)^2}$$

and $$v/c = pc/E$$

Illustrative Examples

1. A train travels through a tunnel at a speed of 3/5 c. The length of the tunnel according to observers at rest with respect to it is 45 m. Find the time the engineer of the train is in the tunnel from the point of view (a) an observer at rest with respect to the tunnel (b) the engineer. (c) Find the length of the tunnel from the point of view of the engineer.

   (a) $\Delta t = \dfrac{\Delta x}{v}$

$$= \frac{45 \text{ m}}{(3/5 \times 3 \times 10^8 \text{ m s}^{-1})}$$

$$= 25 \times 10^{-8} \text{ s}$$

This is an improper time interval because the entrance of the conductor to the tunnel and the conductor's leaving the tunnel happen in two different places for observers in the tunnel frame.

(b) For the conductor, the passing of the front end of the tunnel and the back end of the tunnel happen in the same place. The conductor reads a proper time interval

$$\Delta t' = \Delta t\sqrt{1 - v^2/c^2}$$

$$= \Delta t\sqrt{1 - (3/5)^2}$$

$$= \Delta t\sqrt{1 - 9/25}$$

$$= \Delta t\sqrt{16/25}$$

$$= \Delta t \times 4/5$$

$$= 25 \times 10^{-8} \text{ s} \times 4/5$$

$$= 20 \times 10^{-8} \text{ s}$$

(c) The length of the tunnel according to the conductor is

$$\Delta x' = v\Delta t'$$

$$= (3/5 \times 3 \times 10^8 \text{ m s}^{-1})\ \Delta t'$$

$$= (1.8 \times 10^8 \text{ m s}^{-1})(20 \times 10^{-8} \text{ s})$$

$$= 36 \text{ m}$$

Alternatively, we can find $\Delta x'$ by recognizing that it is an improper length so that

$$\Delta x' = \Delta x\sqrt{1 - v^2/c^2}$$

$$= 45 \text{ m} \times 4/5$$

$$= 36 \text{ m}$$

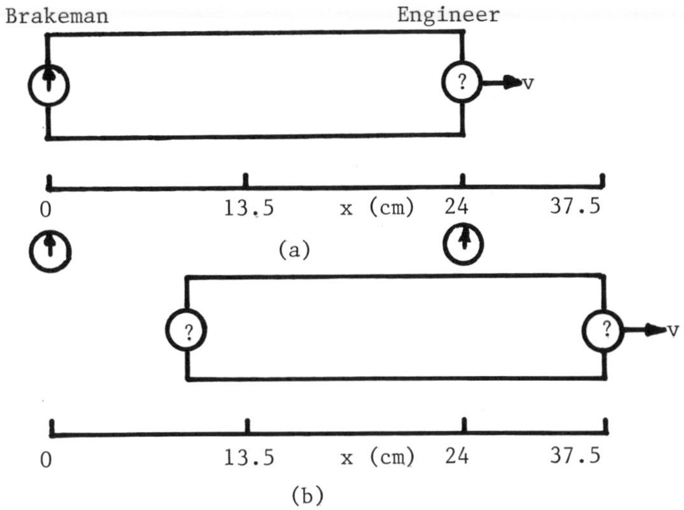

Fig. 11.3

2. Fig. 11.3 represents a train passing a station platform. In Fig. 11.3(a), all platform clocks read $t = 0$ and the position readings are those made by the platform observers. In Fig. 11.3(b), all platform clocks read $t = 7.5 \times 10^{-8}$ s. Find (a) the speed of the train, (b) the length of the train according to observers on the train, (c) the reading on the engineer's clock in Fig. 11.3(a), (d) the reading on the brakeman's clock in Fig. 11.3(b), (e) the reading on the engineer's clock in Fig. 11.3(b), and (f) the length of the platform according to train observers.

(a) $$v = \frac{\Delta x}{\Delta t}$$

$$= \frac{13.5 \text{ m}}{7.5 \times 10^{-8} \text{ s}}$$

$$= 1.8 \times 10^8 \text{ m s}^{-1}$$

$$= 3/5 \times 3 \times 10^8 \text{ m s}^{-1}$$

$$= 3/5\ c$$

(b) $\Delta x' = \dfrac{\Delta x}{\sqrt{1 - v^2/c^2}}$

$= \dfrac{24\text{ m}}{\sqrt{1 - (3/5)^2}}$

$= \dfrac{24\text{ m}}{\sqrt{1 - 9/25}}$

$= \dfrac{24\text{ m}}{\sqrt{16/25}}$

$= \dfrac{24\text{ m}}{4/5}$

$= 24\text{ m} \times 5/4$

$= 30\text{ m}$

(c) The engineer says the platform is shrunk and the point x = 24 m passed him at an earlier time

$t' = \dfrac{-\Delta x'\, v}{c^2}$

$= \dfrac{-30\text{ m} \times 3/5 \times c}{c^2}$

$= \dfrac{-30\text{ m} \times 3/5}{3 \times 10^8\text{ m s}^{-1}}$

$= -6 \times 10^{-8}\text{ s}$

(d) For the brakeman, the passing of x = 0 and x = 24 m happens in the same place so he reads a proper time interval

$\Delta t' = \Delta t\sqrt{1 - v^2/c^2}$

$= 7.5 \times 10^{-8}\text{ s} \times 4/5$

$= 6 \times 10^{-8}\text{ s}$

(e) The engineer's clock also advances by 6×10^{-8} s and in Fig. 11.3(c) reads

$t' = (-6 \times 10^{-8}\text{ s}) + (6 \times 10^{-8}\text{ s})$

$= 0$

(f) The length of the platform according to observers on the train is an improper length

$L = L_0\sqrt{1 - v^2/c^2}$

$= 37.5\text{ m} \times 4/5$

$= 30\text{ m}$

From the point of view of the observers on the train, the length of the platform is the same as the length of the train. For them the brakeman is alongside of the left side of the platform and the engineer is alongside of the right side of the platform at t' = 0.

3. A spaceship is alongside Indianapolis at 12 o'clock noon on July 4th. The speed of the spaceship is 4/5 c. How long by the spaceship's clocks will it take for an observer on the spaceship to receive the light from next year's fireworks?

Observers on the spaceship say that the earth clocks run slow and the time elapsed on their clocks when the fireworks are set off is

$\Delta t' = \dfrac{\Delta t}{\sqrt{1 - v^2/c^2}}$

$\Delta t' = \dfrac{\Delta t}{\sqrt{1 - (4/5)^2}}$

$= \dfrac{\Delta t}{\sqrt{1 - 16/25}}$

$= \dfrac{\Delta t}{\sqrt{9/25}}$

$= \dfrac{1\text{ yr}}{3/5}$

$= 5/3\text{ yr}$

In this time, the earth has traveled a distance 5/3 yr x v. The time for the light to travel this distance and reach the spaceship is

$$t_1' = \frac{5/3 \text{ yr} \times v}{c}$$

$$= 5/3 \text{ yr} \times v/c$$

$$= 5/3 \text{ yr} \times 4/5$$

$$= 4/3 \text{ yr}$$

Observers on the spaceship, therefore, say it takes

$$\Delta t' + t_1' = 5/3 \text{ yr} + 4/3 \text{ yr}$$

$$= 3 \text{ yr}$$

to receive the light from next year's fireworks.

Alternatively, observers on earth say that in 1 yr the spaceship travels a distance v x 1 yr and an additional distance vt, where t is the time for the light to reach the spaceship. The total distance traveled by the light is

$$ct = v \times 1 \text{ yr} + vt$$

or

$$ct - vt = v \times 1 \text{ yr}$$

$$t = \frac{v \times 1 \text{ yr}}{c - v}$$

$$t = \frac{1 \text{ yr}}{(c/v - 1)}$$

$$= \frac{1 \text{ yr}}{(5/4 - 1)}$$

$$= \frac{1 \text{ yr}}{1/4}$$

$$= 4 \text{ yr}$$

The total time measured by observers on earth is 5 yr. They say the spaceship's clocks run slow and the time elapsed on the spaceship's clocks is

$$\Delta t' = \Delta t\sqrt{1 - v^2/c^2}$$

$$= 5 \text{ yr} \sqrt{1 - (4/5)^2}$$

$$= 5 \text{ yr} \times 3/5$$

$$= 3 \text{ yr}$$

Note that both the earth observers and the spaceship observers agree on the reading of the spaceship's clocks when the light reaches these clocks. If they disagreed about the reading on the spaceship's clocks, the theory of relativity would be violated.

4. Trains S' and S" move to the right with velocities 3/5 c and 12/13 c, respectively, relative to a platform S. Find the velocity of S" relative to train S'.

Let the velocity of train S" relative to S be u, the velocity of train S" relative to S' be u', and the relative velocity v of S' and S be 3/5 c. Then

$$u' = \frac{u - v}{1 - uv/c^2}$$

$$= \frac{(12/13 - 3/5)c}{1 - 36/65}$$

$$= \frac{(60/65 - 39/65)c}{(65/65 - 36/65)}$$

$$u' = \frac{21/65\ c}{29/65}$$

$$u' = 21/29\ c$$

5. An object of mass $m = 10^{-20}$ kg is initially at rest. A constant force $F = 10^{-10}$ N acts on the object for time t and it acquires a speed v = 5/13 c. Find (a) t and (b) the displacement s moved by the object in this time.

(a) $Ft = \Delta p$

$$= \frac{m_o \times 5/13\ c}{\sqrt{1 - (5/13)^2}} - 0$$

$$= \frac{m_o \times 5/13\ c}{\sqrt{1 - 25/169}}$$

$$= \frac{m_o \times 5/13\ c}{\sqrt{144/169}}$$

$$= \frac{m_o \times 5/13\ c}{12/13}$$

$$Ft = 5/12\ m_o c$$

or

$$t = \frac{5\ m_o}{12\ F}$$

$$= \frac{5 \times 10^{-20}\ \text{kg} \times 3 \times 10^{8}\ \text{m s}^{-1}}{12 \times 10^{-10}\ \text{N}}$$

$$= 1.25 \times 10^{-2}\ \text{s}$$

(b) The work done

$$W = Fs$$

$$Fs = \Delta E$$

$$= m_o c^2 \left[\frac{1}{\sqrt{1 - v^2/c^2}} - 1 \right]$$

$$= m_o c^2 \left[\frac{1}{12/13} - 1 \right]$$

$$= m_o c^2 \ (13/12 - 12/12)$$

$$= m_o c^2 \times 1/12$$

$$Fs = 1/12\ m_o c^2$$

or

$$s = \frac{m_o c^2}{12\ F}$$

$$= \frac{10^{-20}\ \text{kg} \times 9 \times 10^{16}\ \text{m}^2\ \text{s}^{-2}}{12 \times 10^{-10}\ \text{N}}$$

$$= 7.5 \times 10^{5}\ \text{m}$$

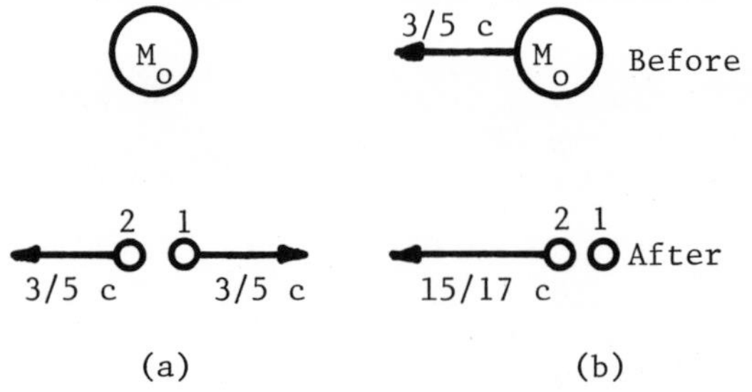

Fig. 11.4

6. An object of rest mass M_o = 5 units at rest decays into two identical particles 1 and 2 of rest mass m_o, as shown in Fig. 11.4(a). The speed of the particles after the decay is 3/5 c. (a) Find m_o. (b) Describe the decay from the point of view of an observer 0', who moves to the right with velocity 3/5 c and show that energy and momentum are conserved in this reference frame.

(a) From conservation of energy

$$M_o c^2 = \frac{m_o c^2}{\sqrt{1 - (3/5)^2}} + \frac{m_o c^2}{\sqrt{1 - (3/5)^2}}$$

$$= \frac{2 m_o c^2}{\sqrt{1 - 9/25}}$$

$$= \frac{2 m_o c^2}{4/5}$$

$$M_o c^2 = 2.5\ m_o c^2$$

or 2 units = m_o

(b) If observer 0 measures a speed u, observer 0', who moves to the right with speed v = 3/5 c measures a speed

$$u' = \frac{u - v}{1 - uv/c^2}$$

For M_o, u = 0 and

$$u' = \frac{0 - 3/5\ c}{1 - 0}$$

$$= -3/5\ c$$

For particle 1, $u = 3/5\ c$ and

$$u' = \frac{3/5\ c - 3/5\ c}{1 - 9/25}$$

$$= 0$$

For particle 2, $u = -3/5\ c$ and

$$u' = \frac{-3/5\ c - 3/5\ c}{1 + 9/25}$$

$$= \frac{-6/5\ c}{34/25}$$

$$= -15/17\ c$$

The decay scheme from the point of view of 0' is shown in Fig. 11 .4(b).

For 0', the energy before

$$E'_b = M_oc^2/\sqrt{1 - (3/5)^2}$$

$$= M_oc^2/(4/5)$$

$$= 5/4\ M_oc^2$$

For 0', the energy after

$$E'_a = m_oc^2 + \frac{m_oc^2}{\sqrt{1 - (15/17)^2}}$$

$$= m_oc^2 + \frac{m_oc^2}{8/17}$$

$$= m_oc^2 + 17/8\ m_oc^2$$

$$E'_a = 25/8\ m_oc^2$$

$$= 25/8\ (M_o/2.5)c^2$$

$$= 5/4\ M_oc^2$$

Notice that $E_b' = E_a'$

For observer 0', the momentum before

$$p_b' = M_o(-3/5\ c)/\sqrt{1 - (3/5)^2}$$

$$= M_o(-3/5\ c)/(4/5)$$

$$= -3/4\ M_oc$$

For observer 0', the momentum after

$$p_a' = m_o(-15/17\ c)/\sqrt{1 - (15/17)^2}$$

$$= m_o(-15/17\ c)/(8/17)$$

$$= -15/8\ m_oc$$

$$= -15/8\ (M_o/2.5)c$$

$$= -3/4\ M_oc$$

and we see that momentum is also conserved for observer 0'.

Problems

1. A spaceship, traveling with a speed of 4/5 c, passes an observer on a reviewing stand on a space platform. The length of the spaceship according to observers on it is 60 m. The length of the spaceship according to observers on the space platform is

   (a) 36 m

   (b) 48 m

   (c) 60 m

2. In Problem 1, the time for the spaceship to pass the observers on the space station is

   (a) 15×10^{-8} s

   (b) 20×10^{-8} s

   (c) 25×10^{-8} s

3. In Problem 1, observers at the front and back of the spaceship, see the observer on the platform pass by in

   (a) 15×10^{-8} s

   (b) 20×10^{-8} s

   (c) 25×10^{-8} s

4. When all the clocks in the platform frame read t = 0, the brakeman on a train, traveling to the right at v = 4/5 c, is alongside x = 0 and the engineer is along side x = 18 m, as read in the platform frame. The length of the train as measured by observers on the train is

(a) 18 m

(b) 22.5 m

(c) 30 m

5. In Problem 4, when the station's clocks read t = 0, the brakeman's clock reads t' = 0. The reading on the engineer's clock is

(a) 0

(b) -4.8×10^{-8} s

(c) -8×10^{-8} s

6. In Problem 4, when the brakeman is alongside x = 18 m, the brakeman's clock reads

(a) 4.5×10^{-8} s

(b) 7.5×10^{-8} s

(c) 12.5×10^{-8} s

7. In Problem 4, when the brakeman is alongside x = 18 m, the engineer's clock reads

(a) 4.5×10^{-8} s

(b) -0.3×10^{-8} s

(c) -3.5×10^{-8} s

8. Spaceships 1 and 2 approach an observer on a space station each at 4/5 the speed of light. Spaceship 1 approaches from the left. Spaceship 2 approaches from the right. The speed of spaceship 2 according to observers on spaceship 1 is

(a) 0

(b) 40/41 c

(c) 8/5 c

9. The work done to bring an object of rest mass m_o from rest to velocity v = 3/5 c is

(a) $m_oc^2/4$

(b) $3/4\ m_oc^2$

(c) $5/4\ m_oc^2$

10. An unidentified particle X decays into two fragments. Each fragment has a rest mass of 4.8 units, and both are moving in the same direction, one with a speed of 7/25 c, the other with a speed of 4/5 c. The total energy of X is

(a) $7.8\ c^2$

(b) $9.6\ c^2$

(c) $13\ c^2$

11. In Problem 10, the momentum of X is

(a) 7.8 c

(b) 13 c

(c) (4.8 x 27c)/25

12. In Problem 10, the velocity of X is

(a) 0

(b) 0.6 c

(c) 0.8 c

(d) 0.28 c

CHAPTER 12. The Duality of Nature

Objectives

You should learn that

1. the energy of a photon

$$E = hf$$

and its momentum

$$p = h/\lambda$$

$$= hf/c$$

where h is Planck's constant and f and λ are the frequency and wavelength of the photon, respectively.

2. when a photon is absorbed by a photosensitive surface, part of its energy goes into ejecting the electron from the surface and the remaining energy goes into the kinetic energy of the electron

3. the scattered photon in the Compton effect has a larger wavelength than the incident photon and its wavelength depends on the angle of scattering

4. conservation of momentum and energy can be used to find the velocity of the electron scattered in the Compton effect

5. a material particle has wave properties with wavelength

$$\lambda = h/mv$$

where m and v equal the mass and velocity, respectively, of the particle

6. the wave properties of material particles means that we can no longer measure the momentum and position of a particle with absolute certainty

7. when we try to measure the position of a particle and its momentum simultaneously the uncertainty in position Δx and the uncertainty in momentum Δp are related through the expression

$$\Delta x \, \Delta p = h/2\pi$$

8. you can ignore the particle properties of light for electromagnetic waves of very long wavelength

9. you can ignore the wave properties of material particles for objects of large mass

10. when an event can occur in alternative ways and the ways are not distinguishable, its probability amplitude is the sum of the amplitudes for the alternatives,

$$\vec{A} = \vec{A}_1 + \vec{A}_2$$

and the probability

$$P'_{12} = A^2$$

11. when an event occurs in such a way that the alternatives are distinguishable,

$$P_{12} = A_1^{\,2} + A_2^{\,2}$$

$$= P_1 + P_2$$

Trouble Spots

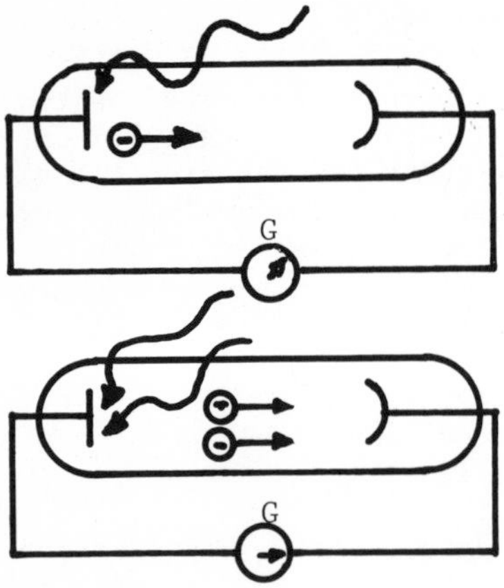

Fig. 12.1

In the photoelectric effect, the reading of the galvanometer is a measure of the number of electrons released from the photosensitive surface. In Fig. 12.1, when the intensity of the light is increased, the number of electrons emitted from the photosensitive surface is increased.

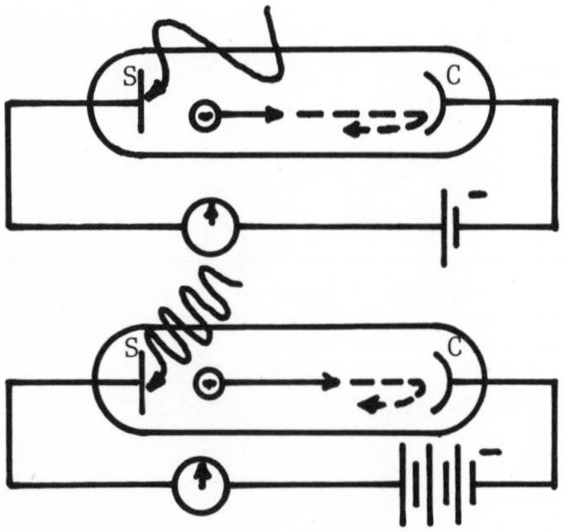

Fig. 12.2

The maximum kinetic energy of the electrons is found by determing what potential difference is needed to bring the galvanometer reading to zero. In Fig. 12.2(b), a greater number of batteries, and therefore a greater V_o, is needed than in Fig. 12.2(a) because the frequency of the light in Fig. 12.2(b) is greater than the frequency of the incident light in Fig. 12.2(a). The electrons leave the photosensitive surface with a kinetic energy

$$KE = 1/2\ m\ v^2_{max}$$

They are brought to rest by the negatively charged collector. The change in kinetic energy

$$\Delta KE = 1/2\ mv^2_{max} - 0$$
$$= 1/2\ mv^2_{max}$$

In passing from the photosensitive surface s to the collector c, the electron gains potential energy

$$PE_s - PE_c = eV_o$$

From conservation of energy,

$$1/2\ mv^2_{max} = eV_o$$

In the photoelectric effect, the photon ceases to exist. In the Compton effect, the photon exists after interacting with an electron but its wavelength is increased. Momentum is conserved in the photoelectric effect because the photosensitive surface recoils. Because of its large mass, the surface acquires very little of the photon energy. For this reason, we ignore the energy transferred to the surface in the photoelectric effect.

When the energy of a photon is doubled, the maximum kinetic energy acquired by the photoelectron is <u>not</u> doubled. The kinetic energy of the electron is <u>not</u> directly proportional to the energy of the incident photon because some of the energy of the photon is used to remove the electron from the surface. In other words,

$$hf = hf_o + 1/2\ mv^2$$

while

$$hf \neq 1/2\ mv^2$$

where hf_o is the energy required to remove the electron from the surface.

In working problems on the Compton effect, remember that momentum is a vector quantity, while energy is a scalar quantity. For example if the photon is scattered through 180°, from conservation of energy

$$hf + m_oc^2 = hf' + mc^2$$

From conservation of momentum

$$hf/c = -hf'/c + mv$$

where f and f' are the frequency of the incident and scattered photon, respectively, and m_o and m are the mass of the electron when it is at rest and when it moves with speed v after the collision, respectively. Note that the minus sign before hf'/c indicates that the momentum of the scattered photon is opposite in direction to that of the incident photon, while the plus sign before mv indicates that the momentum of the scattered electron is in the same direction as that of the incident photon, hf/c.

The equation, $\lambda = h/mv$, indicates that we are dealing with a material particle of mass m that moves with speed $v \neq c$. For a photon

$$E = hc/\lambda$$

$$= 1.24 \times 10^{-6} \text{ eV}/\lambda$$

where λ is in meters. Since the speed of light appears in the above equation, the energy of a material particle

$$E \neq 1.24 \times 10^{-6} \text{ eV}/\lambda$$

The total energy of a material particle of rest mass m_o and momentum p is

$$E = \sqrt{(pc)^2 + (m_oc^2)}$$

The uncertainty principle states that the product of the uncertainty in position Δx and the uncertainty in momentum Δp_x is greater or equal to $h/2\pi$ when you try to measure the position and momentum of the object simultaneously. You can measure the position of an object or its momentum with absolute certainty, but then you lose all information about the other quantity.

The wavelength and frequency of a photon are related through the equation

$$\lambda \times f = c$$

When you increase the wavelength of a photon, you decrease its frequency.

Illustrative Examples

1. A photon of wavelength $\lambda = 2.48 \times 10^{-7}$ m is incident upon a surface. The energy required to release an electron from the surface is 2.0 eV. (a) What is the threshold frequency? (b) What is the maximum kinetic energy of the electrons released? (c) If light of wavelength $\lambda = 3.1 \times 10^{-7}$ m falls on the surface, what stopping potential V_o is required to reduce the galvanometer reading to zero?

(a) The threshold wavelength λ_o, where $E = hf_o$ in eV is

$$\lambda_o = \frac{1.24 \times 10^{-6}}{2.0} \text{ m}$$

$$= 6.2 \times 10^{-7} \text{ m}$$

Then

$$f_o = \frac{c}{\lambda_o}$$

$$= \frac{3 \times 10^8 \text{ m s}^{-1}}{6.2 \times 10^{-7} \text{ m}}$$

$$= 4.8 \times 10^{14} \text{ s}^{-1}$$

(b) A photon of wavelength $\lambda = 2.48 \times 10^{-7}$ m has an energy in eV of

$$E = \frac{1.24 \times 10^{-6} \text{ eV}}{2.48 \times 10^{-6}}$$

$$= 5 \text{ eV}$$

The energy of the photon goes into the energy to remove the electron from the surface, hf_o, and the kinetic energy of the electron, that is,

$$E = hf_o + (KE)_{max}$$

$$5.0 \ \text{eV} = 2.0 \ \text{ev} + (KE)_{max}$$

$$3.0 \ \text{eV} = (KE)_{max}$$

(c) A photon of wavelength $\lambda = 3.1 \times 10^{-7}$ m has an energy in eV of

$$E = \frac{1.24 \times 10^{-6} \ \text{eV}}{3.1 \times 10^{-7}}$$

$$= 4.0 \ \text{eV}$$

For this case,

$$KE_{max} = 4.0 \ \text{eV} - 2.0 \ \text{eV}$$

$$= 2.0 \ \text{eV}$$

The electron loses this kinetic energy by gaining a potential energy

$$PE = 2.0 \ \text{eV}$$

$$= 2.0 \times 1.6 \times 10^{-19} \ \text{J}$$

Since

$$PE = eV_o,$$

$$eV_o = 2.0 \times 1.6 \times 10^{-19} \ \text{J}$$

Because the charge of the electron $e = 1.6 \times 10^{-19}$ C,

$$1.6 \times 10^{-19} \text{C} \ V_o = 2.0 \times 1.6 \times 10^{-19} \ \text{J}$$

$$V_o = 2.0 \ \text{J C}^{-1}$$

$$= 2.0 \ \text{V}$$

2. A photon of frequency f collides with a stationary electron of rest mass m_o and is scattered straight backwards with a lower frequency f_o. In the process the electron gains a speed $v = 5/13c$. (a) How much kinetic energy did the electron acquire? What is its momentum after the collision? (b) Write down two equations involving the unknowns f and f' that describe conservation of energy and momentum in this collision. (c) Solve the equations for f and f', and check that your solution agrees with the Compton formula for wavelength increase.

(a) The total energy of the electron after the collision

$$E = mc^2$$

$$= \frac{m_o c^2}{\sqrt{1 - (5/13)^2}}$$

$$= \frac{m_o c^2}{\sqrt{144/169}}$$

$$= \frac{m_o c^2}{12/13}$$

$$= 13/12 \ m_o c^2$$

The kinetic energy of the electron

$$KE = E - m_o c^2$$

$$= 13/12 \ m_o c^2 - m_o c^2$$

$$= m_o c^2/12$$

The momentum of the electron

$$p = \frac{m_o v}{\sqrt{1 - v^2/c^2}}$$

$$= \frac{m_o \times 5/13 \ c}{\sqrt{1 - (5/13)^2}}$$

$$= \frac{m_o \times 5/13 \ c}{12/13}$$

$$= 5/12 \ m_o c$$

(b) From conservation of energy

$$hf + m_o c^2 = hf' + mc^2$$

or

$hf - hf' = mc^2 - m_oc^2 = KE$

$$hf - hf' = m_oc^2/12 \qquad (1)$$

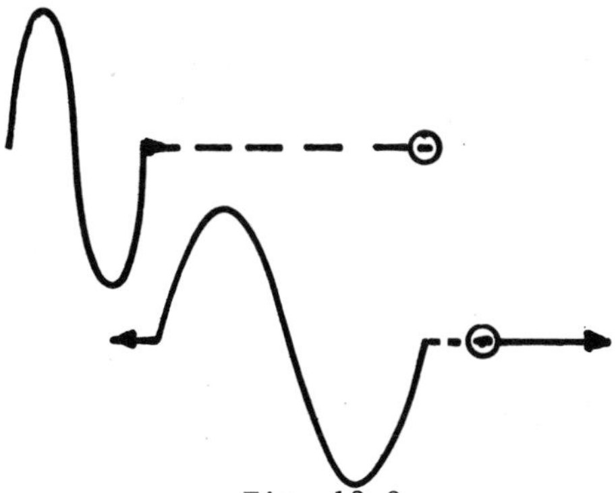

Fig. 12.3

The collision is shown in Fig. 12.3. Assigning a positive sign for momentum to the right, we write for conservation of momentum,

$$hf/c = -hf'/c + p$$

$$hf/c = -hf'/c + 5m_oc/12$$

or

$$hf + hf' = 5m_oc^2/12 \qquad (2)$$

(c) Adding Equation 1 and Equation 2, we find

$$2hf = 6m_oc^2/12$$

$$f = m_oc^2/4h$$

Subtracting Equation 1 from Equation 2, we get

$$2hf' = 4m_oc^2/12$$

$$f' = m_oc^2/6h$$

In general, wavelength = c/frequency so we find

$$\lambda' = c/f'$$

$$= c/(m_oc^2/6h)$$

$$= 6h/m_oc$$

and $\lambda = c/f$

$$= c/(m_oc^2/4h)$$

$$= 4h/m_oc$$

$$\lambda' - \lambda = 6h/m_oc - 4h/m_oc$$

$$= 2h/m_oc$$

The Compton formula for change in wavelength is

$$\lambda' - \lambda = h/m_oc\ (1 - \cos\theta)$$

For backward scattering, $\theta = 180^o$. Since $\cos 180^o = -1$

$$\lambda' - \lambda = h/m_oc(1 - (-1))$$

$$= 2h/m_oc$$

3. Compare the total energy in electron volts of an electron and photon if the wavelength of each is 3×10^{-12} m.

For the electron,

$$p = h/\lambda$$

$$= \frac{6.63 \times 10^{-34}\ \text{J-s}}{3 \times 10^{-12}\ \text{m}}$$

$$= 2.21 \times 10^{-22}\ \text{kg-m s}^{-1}$$

Its total energy

$$E = \sqrt{(pc)^2 + (m_oc^2)^2}$$

where $pc = 2.21 \times 10^{-22}$ kg-m s^{-1} x 3 x 10^8 m s^{-1} = 6.63×10^{-14} J. Since we are requested to give our answer in electron volts, we convert pc in joules to pc in eV.

$$pc = \frac{6.63 \times 10^{-14}\ \text{J}}{1.6 \times 10^{-19}\ \text{J-eV}^{-1}}$$

$$= 0.41 \times 10^6\ \text{eV}$$

And the rest mass energy in electron volts

$$m_oc^2 = \frac{9.1 \times 10^{-31} kg \times 9 \times 10^{16} m^2\ s^{-2}}{1.6 \times 10^{-19}\ J\text{-}eV^{-1}}$$

$$= 0.51 \times 10^6\ eV$$

$$E = \sqrt{(pc)^2 + (m_oc^2)^2}$$

$$= \sqrt{(0.41)^2 + (0.51)^2} \times 10^6\ eV$$

$$= 0.66 \times 10^6\ eV$$

The energy of the photon in eV is

$$E = \frac{1.24 \times 10^{-6}\ eV}{3 \times 10^{-12}}$$

$$= 0.41 \times 10^6\ eV$$

4. A particle of mass $m = 10^{-20}$ kg travels with a speed of $v = 10^6\ m\ s^{-1}$. Find the wavelength of the particle. Comment on the practicality of discovering its wave properties.

$$\lambda = \frac{h}{mv}$$

$$= \frac{6.63 \times 10^{-34}\ J\text{-}s}{10^{-20}\ kg \times 10^6\ m\ s^{-1}}$$

$$= 6.63 \times 10^{-20}\ m$$

We found it difficult to detect interference and diffraction effects of x-rays with wavelength λ approximately equal to 10^{-10} m. It is much more difficult to detect the wave properties for a wavelength of 6.63×10^{-20} m because the smaller the wavelength the greater is the difficulty in detecting wave properties.

5. The particles, described in Problem 4 pass through a slit of width $w = 10^{-7}$ m and hit a screen at a distance $L = 1$ m. What is the width of the central maximum.

The width of the central maximum

$$2D_1 = 2\lambda L/w$$

$$= \frac{2 \times 6.63 \times 10^{-20}\ m}{10^{-7}\ m}$$

$$= 13.26 \times 10^{-13}\ m$$

Since the width of the central maximum is much less than the width of the slit, the diffraction effects are impossible to detect.

6. An electron is confined to an atom within a distance $\Delta x = 10^{-10}$ m. (a) Find the uncertainty in the momentum of the electron. (b) Assuming the momentum of the electron is, at least, as great as the minimum uncertainty in the momentum, find the kinetic energy of the electron. This is a non-relativistic problem.

(a) $\Delta p_x \Delta x \gtrsim h/\ 2\pi$

$$\gtrsim 1.06 \times 10^{-34}\ J\text{-}s$$

$$\Delta p_x \gtrsim 1.06 \times 10^{-34}\ J\text{-}s/\Delta x$$

$$\gtrsim 1.06 \times 10^{-34}\ J\text{-}s/10^{-10}\ m$$

$$\gtrsim 1.06 \times 10^{-24}\ kg\text{-}m\ s^{-1}$$

As a lower limit $p_x = 1.06 \times 10^{-24}\ kg\text{-}m\ s^{-1}$.

(b) The kinetic energy of the electron

$$KE = 1/2\ mv^2$$

$$= (mv)^2/2m$$

$$= p_x^2/2m$$

$$= \frac{(1.06 \times 10^{-24}\ kg\text{-}m\ s^{-1})^2}{2 \times 9.1 \times 10^{-31}\ kg}$$

$$= 0.06 \times 10^{-17}\ J$$

$$= \frac{0.06 \times 10^{-17}\ J}{1.6 \times 10^{-19}\ J\text{-}ev^{-1}}$$

$$= 3.7\ eV$$

Since the kinetic energy of the electron (3.7 eV) is much less than its rest mass energy (0.51×10^6 eV), we are justified in using the classical expression for the kinetic energy.

7. Electrons pass through a double slit arrangement. When slit S_1 is open, 1600 electrons arrive at the center of the

screen. When slit S_2 is open, 1600 electrons arrive at the center of the screen. How many electrons arrive at the center of the screen when both slits are open?

The events can happen in alternative ways. We take the probability P_1 of the electrons arriving at the center of the screen to be 1600 when slit S_1 is open and the probability $P_2 = 1600$ when slit S_2 is open.

Since $P_1 = A_1^2$,

$$A_1^2 = 1600$$

$$A_1 = 40$$

and $P_2 = A_2^2$ so,

$$A_2 = 40$$

Since the electron waves are in phase at the center of the screen,

$$A = A_1 + A_2$$

$$= 40 + 40$$

$$= 80$$

The probability of the electrons arriving at the center of the screen when both S_1 and S_2 are open is

$$P'_{12} = A^2$$

$$= (80)^2$$

$$= 6400$$

Note that when the events can happen in alternative ways, and the ways are not distinguishable,

$$P'_{12} \neq A_1^2 + A_2^2$$

Problems

1. Photon 1 has frequency f. Photon 2 has frequency 2f. The ratio of the energy of photon 1 to photon 2 is

(a) 1/2

(b) 1

(c) 2

2. In Problem 1, the momentum of photon 1 to the momentum of photon 2 is

(a) 1/2

(b) 1

(c) 2

3. A photon of wavelength $\lambda = 2.48 \times 10^{-7}$ m hits a photosensitive surface. The maximum kinetic energy of the electron released from the surface is

(a) 0

(b) Less than 5.0 eV

(c) 5.0 eV

4. The intensity of light falling on a photosensitive surface is I. The intensity is increased to 2I. Which of the following is true?

(a) it takes twice the voltage to stop the electrons for intensity 2I

(b) the electrons are emitted earlier by a factor of two

(c) the number of electrons is increased by a factor of two

5. Photons are incident upon free electrons that are essentially at rest. After the photons are scattered by the electrons, the photons

(a) do not exist

(b) have a greater frequency

(c) have a greater wavelength

(d) none of the above

6. Photons of wavelength $\lambda = 6.2 \times 10^{-7}$ m are incident upon a surface. The stopping potential is set so that the galvanometer reads zero. When light of wavelength 3.1×10^{-7} m is incident upon the surface the maximum kinetic energy of the electrons is

(a) 0

(b) 2 eV

(c) 4 eV

(d) 6 eV

Fig. 12.4

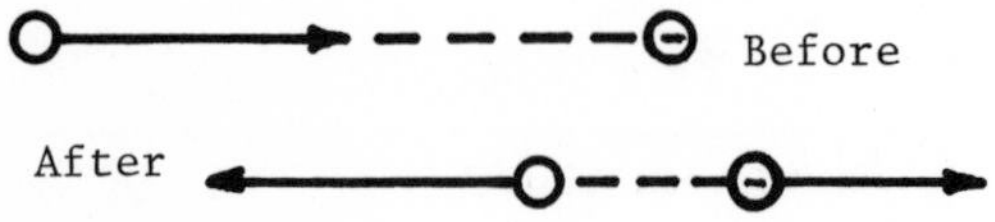

7. A photon of energy $1/3\ m_oc^2$ is incident upon an electron of rest mass m_o. The scattered photon moves opposite in direction to the incident photon (Fig. 12.4). The frequency of the scattered photon is

(a) $m_oc^2/3h$

(b) $m_oc^2/4h$

(c) $m_oc^2/5h$

8. In Problem 7, the velocity of the scattered electron is

(a) 3/5 c

(b) 8/17 c

(c) 4/5 c

(d) 15/17 c

9. A beam of electrons is accelerated through a potential difference of 100 V. The electron wavelength is

(a) 1.24×10^{-4} m

(b) 1.24×10^{-6} m

(c) 1.24×10^{-8} m

(d) 1.24×10^{-10} m

10. When only slit S_1 is open 100 electrons reach a detector at P. When slit S_2 is open 900 electrons reach the detector. The number of electrons that reach P when both S_1 and S_2 are open if the electron waves arrive in phase is

(a) 400

(b) 1000

(c) 1600

11. An electron and a photon both have a wavelength equal to 10^{-12} m. Which of the following is true?

(a) both the electron and the photon have the same momentum

(b) both the electron and the photon have the same total energy

(c) both the electron and the photon have the same kinetic energy

12. A baseball of mass 10^{-1} kg and speed 20 m can best be described

(a) as a particle

(b) as a wave

Sample Quiz for Chapters 11 and 12

1. A particle passes through a tube at a speed v = 7/25 c. The length of the tube, as measured by observers at rest with respect to it is 21 m. The time for the particle to pass through the tube according to an observer at rest with respect to the particle is

   (a) 75 s

   (b) 24×10^{-8} s

   (c) 25×10^{-8} s

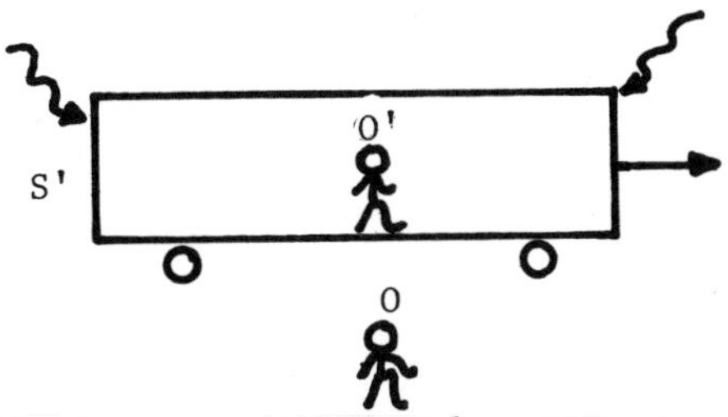

Fig. 1

2. Assume that when train S' passes a station, two bolts of lightning strike the ends of the train simultaneously, according to observer O on the platform (Fig. 1). According to observer O' on the train,

   (a) both light pulses reach him simultaneously

   (b) the pulse from the rear end of the train reaches him first

   (c) the pulse from the front end of the train reaches him first

3. In Problem 2, the train observer explains the report of observer O by saying

   (a) the train moves to the right

   (b) the platform moves to the left

4. Observers in reference frame S', moving to the right with speed v = 4/5 c relative to reference frame S, measures the speed of a particle to be 3/5 c to the left. The velocity of the particle according to observers in S is

   (a) 1/5 c to the right

   (b) 5/13 c to the right

   (c) 35/37 c to the left

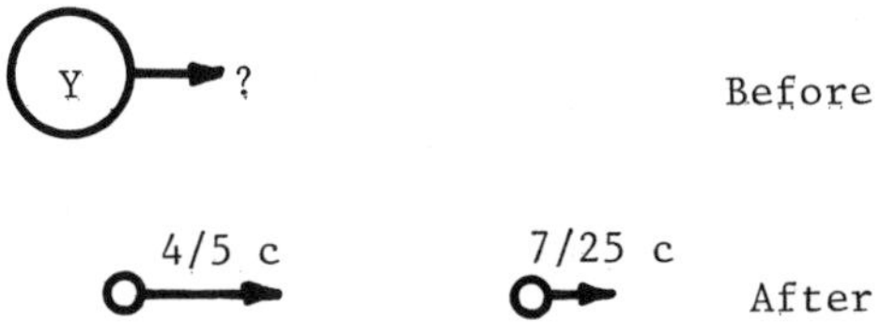

Fig. 2

5. Particle Y decays into two particles of rest mass 12 units. Both particles move in the same direction after the decay, one with speed 7/25 c and the other with speed 4/5 c (Fig. 2). The total energy of Y is

   (a) $19.5\ c^2$

   (b) $24\ c^2$

   (c) $32.5\ c^2$

6. In Problem 5, the velocity of particle Y is

   (a) 4/5 c

   (b) 13/25 c

   (c) 3/5 c

7. If you double the frequency of the incident light hitting a photosensitive surface,

   (a) the number of electrons released from the surface is doubled

   (b) the maximum kinetic energy of the electrons released is doubled

(c) the maximum kinetic energy of the electrons released is greater, but not doubled

8. The wavelength of a photon is 3.1×10^{-7} m. The energy of the photon is

(a) 2.0 eV

(b) 3.0 eV

(c) 4.0 eV

9. To remove an electron from a surface, you must supply an energy of 2.0 eV. The threshold wavelength is

(a) 12.4×10^{-7} m

(b) 6.2×10^{-7} m

(c) 4.1×10^{-7} m

10. When the photon of Problem 8 falls on the surface of Problem 9, the maximum kinetic energy of the electrons is

(a) 0

(b) 1.0 eV

(c) 2.0 eV

(d) 3.0 eV

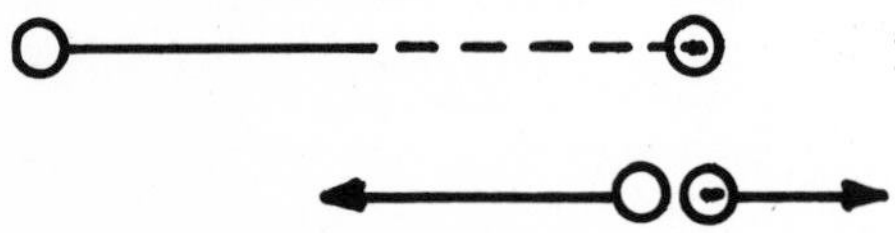

Fig. 3

11. A photon of energy $m_0c^2/2$ is incident upon a stationary electron of rest mass m_0. The scattered photon moves in the opposite direction to the incident photon (Fig. 3). The frequency of the scattered photon is

(a) $m_0c^2/16h$

(b) $m_0c^2/4h$

(c) $m_0c^2/2h$

12. In Problem 11, the velocity of the scattered electron is

(a) 8/17 c

(b) 3/5 c

(c) 4/5 c

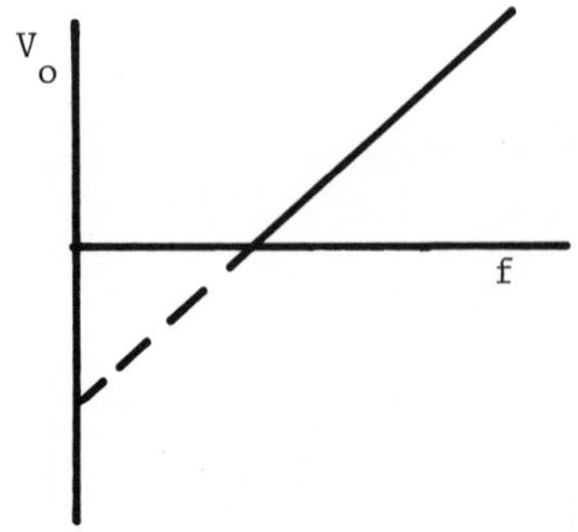

Fig. 4

13. Figure 4 is a plot of the stopping potential V_o as a function of frequency f. The intercept on the V_o- axis is

(a) f_o

(b) $1/2\ mv^2_{max}$

(c) hf_o

(d) hf_o/e

14. If Figure 4 were plotted for another photosensitive surface, it would have

(a) the same intercept on the f-axis

(b) the same intercept on the V_o axis

(c) the same slope

CHAPTER 13. Quantum Systems and the Structure of Atoms

Objectives

You should learn

1. to identify the energy of a photon as the difference in energy levels through which the excited atom falls when it emits the photon

2. that the nth level for hydrogen has energy

$$E_n = - 13.6 \text{ eV}/n^2$$

3. how the wave properties of a particle leads to quantization of energy levels for bound systems

4. to differentiate between the wavelength of a particle in a given level and the wavelength of light emitted when the particle goes from a higher energy level to a lower energy level

5. the boundary conditions that lead to discrete energy levels for an electron in the one-dimensional infinite well

6. how the dependence of the potential energy function on distance for the hydrogen atom leads to a variation in wavelength within a given energy level

7. when an atom is hydrogen-like

8. that the energy for hydrogen-like atoms depends on the effective charge "seen" by the valence electron

9. how the potential energy depends on the quantum numbers n, ℓ, m_ℓ, and m_s

10. how to apply the Pauli Exclusion Principle to find the electron configuration of atoms

Trouble Spots

Fig. 13.1

An energy level diagram for H is shown in Fig. 13.1. The energy for the ground state is -13.6 eV and the energy of the first excited state (n = 2) is -3.4 eV. When a hydrogen atom drops from n = 2 to n = 1, the energy loss by the atom is

$$E_2 - E_1 = -3.4 \text{ eV} - (-13.6 \text{ eV})$$

$$= 10.2 \text{ eV}$$

The wavelength in meters of the photon emitted when the atom goes from n = 2 to n = 1 is

$$\lambda = \frac{1.24 \times 10^{-6}}{10.2} \text{ m}$$

$$= 1.22 \times 10^{-7} \text{ m}$$

Never use the absolute value of an energy level to find the energy of the emitted photon. For example,

$$\lambda \neq \frac{1.24 \times 10^{-6}}{3.4} \text{ m}$$

nor

$$\lambda \neq \frac{1.24 \times 10^{-6}}{13.6} \text{ m}$$

The energy of the nth level for hydrogen is

$$E_n = -13.6 \text{ eV}/n^2$$

This is not the correct expression for the energy of any other atom. For singly ionized helium or doubly ionized lithium,

$$E_n = -13.6\ Z^2 \text{ eV}/n^2$$

where Z = 2 for helium and Z = 3 for lithium.

For hydrogen-like atoms (atoms that have a single valence electron) such as sodium,

$$E_n = -13.6\ (Z_{eff})^2/n^2$$

where Z_{eff} is the effective nuclear charge "seen" by the valence electron. The effective charge varies with the value of n and the value of ℓ.

Contrary to classical theory, the orbital angular momentum of an atom can be zero. The orbital angular quantum number ℓ for a principal quantum number n can take on the values

$$0, 1, 2 \ldots\ldots (n - 1)$$

The possible values of m_ℓ are

$$-\ell, (-\ell+1) \ldots\ldots 0, 1, 2 \ldots\ldots \ell$$

The quantum number m_s can be either -1/2 or 1/2.

A bombarding electron can excite an atom to any energy level, provided it has enough energy, because the electron exists after the interaction and can have angular momentum. Angular momentum can be conserved because the angular momentum of the electron and the atom before the interaction equals the angular momentum of the electron and the atom after the interaction. When a photon with angular momentum $h/2\pi$ interacts with an atom and is absorbed, the change in energy level is such that $\Delta\ell = 1$ so that angular momentum is conserved.

A discharge tube contains many atoms. Different atoms can be excited to different levels and return to the ground state by various routes. Thus lines of many series are observed.

Illustrative Examples

1. Hydrogen gas in its ground state is bombarded with electrons of energy 12.1 eV. What are the wavelengths of light that can be emitted?

In general, for hydrogen

$$E_n = -13.6 \text{ eV}/n^2$$

For n = 1,

$$E_1 = -13.6 \text{ eV}/1$$

$$= -13.6 \text{ eV}$$

For n = 2,

$$E_2 = -13.6 \text{ eV}/4$$

$$= -3.4 \text{ eV}$$

For n = 3,

$$E_3 = -13.6 \text{ eV}/9$$

$$= -1.5 \text{ eV}$$

To take the electron from n = 1 to n = 3 requires an energy

$$E_3 - E_2 = -1.5 \text{ eV} - (-13.6 \text{ eV})$$

$= 12.1$ eV

Thus the bombarding electron excites the atom to $n = 3$. The atom can return to the ground state by going from $n = 3$ to $n = 2$ and then from $n = 2$ to $n = 1$ or it can go directly from $n = 3$ to $n = 1$. For a transition from $n = 3$ to $n = 2$, the wavelength of the light in meters is

$$\lambda = \frac{1.24 \times 10^{-6}}{E_3 - E_2} \text{ m}$$

$$= \frac{1.24 \times 10^{-6}}{-1.5 - (-3.4)} \text{ m}$$

$$= \frac{1.24 \times 10^{-6}}{1.9} \text{ m}$$

$$= 6.53 \times 10^{-7} \text{ m}$$

For a transition from $n = 2$ to $n = 1$,

$$\lambda = \frac{1.24 \times 10^{-6}}{E_2 - E_1} \text{ m}$$

$$= \frac{1.24 \times 10^{-6}}{-3.4 - (-13.6)} \text{ m}$$

$$= 1.22 \times 10^{-7} \text{ m}$$

For a transition from $n = 3$ to $n = 1$,

$$\lambda = \frac{1.24 \times 10^{-6}}{E_3 - E_1} \text{ m}$$

$$= \frac{1.24 \times 10^{-6}}{-1.9 - (-13.6)} \text{ m}$$

$$= 1.02 \times 10^{-7} \text{ m}$$

2. Suppose a hydrogen atom, at rest in interstellar space, absorbs a photon that excites it to $n = 3$. Find what velocity the hydrogen atom acquires in this process. The mass m of the hydrogen atom is 1.67×10^{-27} kg.

From Problem 1, we know that $E_3 - E_1 = 12.1$ eV and that when the atom goes from $n = 3$ to $n = 1$, it emits a photon with $\lambda = 1.02 \times 10^{-7}$ m. Thus to excite the atom from $n = 1$ to $n = 3$, the photon has a wavelength of 1.02×10^{-7} m. The momentum of this photon is

$$p = h/\lambda$$

$$= \frac{6.63 \times 10^{-34} \text{ J-s}}{1.02 \times 10^{-7} \text{ m}}$$

$$= 6.5 \times 10^{-27} \text{ kg-m s}^{-1}$$

To conserve momentum, the atom must recoil with this momentum. If v is the speed of the hydrogen atom,

$$v = p/m$$

$$= \frac{6.5 \times 10^{-27} \text{ kg-m s}^{-1}}{1.67 \times 10^{-27} \text{ kg}}$$

$$= 3.89 \text{ m s}^{-1}$$

3. Suppose that a new type of atom has been discovered in which a negatively charged particle has the same charge as the electron but a different mass. The longest wavelength that this atom can absorb when it is in the ground state is 64×10^{-10} m. (d) Determine one other wavelength that the atom can absorb. (b) What is the ratio of its mass m' to the electron mass m?

(a) Since the atom is normally in $n = 1$, the longest wavelength (shortest frequency) that the atom can absorb (see Fig. 13.1) takes the atom from $n = 1$ to $n = 2$. From the Bohr theory of the atom

$$E_2 - E_1 = 2\pi^2 m' k_e e^4/h^2 (1/1 - 1/4)$$

or

$$2\pi^2 m' k_e e^4/h^2 (3/4) = \frac{1.24 \times 10^{-6}}{64 \times 10^{-10} \text{ m}} \quad (1)$$

For a transition from $n = 1$ to $n = 3$,

$$E_3 - E_1 = 2\pi^2 m' k_e e^4/h^2 (1/1 - 1/9)$$

or

$$2\pi^2 m' k_e e^4/h^2 (8/9) = \frac{1.24 \times 10^{-6}}{\lambda} \quad (2)$$

Dividing Eq. 1 by Eq. 2, we find

$$\frac{3/4}{8/9} = \frac{\lambda}{64 \times 10^{-10}\ \text{m}}$$

$$\lambda = 54 \times 10^{-10}\ \text{m}$$

(b) For hydrogen $E_2 - E_1 = 10.2$ eV and the wavelength of the photon to excite the hydrogen atom from n = 1 to n = 2 is

$$\lambda = \frac{1.24 \times 10^{-6}}{10.2}\ \text{m}$$

$$= 1.22 \times 10^{-7}\ \text{m}$$

or

$$2\pi^2 mk_e e^4/h^2(3/4) = \frac{1.24 \times 10^{-6}}{1.22 \times 10^{-7}\ \text{m}} \qquad (3)$$

Dividing Eq. 1 by Eq. 3, we find

$$\frac{m'}{m} = \frac{1.22 \times 10^{-7}}{64 \times 10^{-10}}$$

$$= 19$$

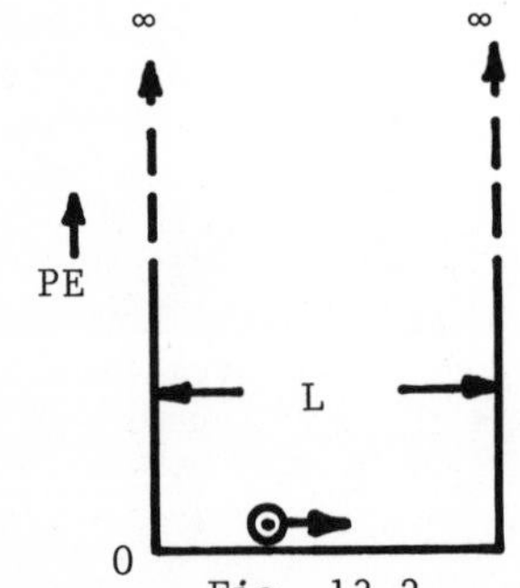

Fig. 13.2

4. An electron moves back and forth in a one-dimensional well of length L = 10^{-10} m (Fig. 13.2). The sides of the well are infinitely high. (a) What is the lowest energy the electron can have? (b) What wavelength of light is emitted in the transition from n = 2 to n = 1?

(a) For a particle of mass m in a one-dimensional well of length L with infinitely high sides, the possible energies are

$$E_n = \frac{n^2h^2}{8mL^2}$$

Since the smallest allowed value of n is 1, the lowest energy is

$$E_1 = \frac{1^2 \times 6.63 \times 10^{-34}\ \text{J-s}}{8 \times 9.1 \times 10^{-31}\ \text{kg} \times 10^{-20}\ \text{m}^2}$$

$$= 6.03 \times 10^{-18}\ \text{J}$$

$$= \frac{6.03 \times 10^{-18}\ \text{J}}{1.6 \times 10^{-19}\ \text{J-eV}^{-1}}$$

$$= 37.7\ \text{eV}$$

(b) $$E_2 - E_1 = 4h^2/8mL^2 - h^2/8mL^2$$

$$= 3\ (h^2/8mL^2)$$

$$= 3\ (37.7\ \text{eV})$$

$$= 113.1\ \text{eV}$$

$$= \frac{1.24 \times 10^{-6}}{113.1}\ \text{m}$$

$$= 1.10 \times 10^{-8}\ \text{m}$$

5. Assume the ground state for the outer electron of a certain atom is n = 3 and $\ell = 0$ with $Z_{eff} = 3.2$. Assume the next highest level for this atom is n = 3 and $\ell = 1$ with $Z_{eff} = 1.6$. What is the wavelength of the photon emitted in a transition from n = 3, ℓ to the ground state?

In general,

$$E_n = -13.6\ Z^2_{eff}\ \text{eV}/n^2$$

For n = 3, $\ell = 0$,

$$E_3 = \frac{-13.6\ (1.6)^2\ \text{eV}}{9}$$

$$= -3.87\ \text{eV}$$

For n = 3, $\ell = 0$,

$$E_3 = \frac{-13.6 \times (3.2)^2 \text{ eV}}{9}$$

$$= -15.48$$

When the atom drops from $n = 3$, $\ell = 1$ to $n = 3$, $\ell = 0$, the difference in energy is

$$\Delta E = -3.87 \text{ eV} - (-15.48 \text{ eV})$$

$$= 11.61 \text{ eV}$$

and the wavelength of the light emitted is

$$= \frac{1.24 \times 10^{-6}}{11.61} \text{ m}$$

$$= 1.07 \times 10^{-7} \text{ m}$$

6. The electron configuration of potassium ($Z = 19$) is

$$1s^2 2s^2 2p^6 3s^2 3p^6 4s^1$$

The electron configuration of calcium ($Z = 20$) is

$$1s^2 2s^2 2p^6 3s^2 3p^6 4s^2$$

In the next element, scandium ($Z = 21$) the 3 d states start to fill up. How many electrons will it take to completely fill up the 3d state.

The d state corresponds to $\ell = 2$. For $\ell = 2$, the possible values of m_ℓ are

2, 1, 0, -1, 2

For each value of m_ℓ, the possible values of m_s are

-1/2 and + 1/2

Therefore, there can be (5 x 2) or 10 electrons in the 3d state. Since scandium adds one electron to the 3d state, the 3d state can take 9 additional electrons.

Problems

1. Electrons bombard hydrogen atoms in their ground state. If the initial energy of the electrons is 10.5 eV, after interacting with the hydrogen atoms the electrons have an energy of

   (a) 0 eV

   (b) 0.3 eV

   (c) 10.5 eV

   (d) a continuous range of energies

2. Hydrogen atoms are excited from their ground state to $n = 4$. When the atoms return to their ground state, the number of lines that appear in the spectrum is

   (a) 1

   (b) 2

   (c) 3

   (d) 4

   (e) 5

   (f) 6

   (g) 7

3. When singly ionized helium ($Z = 2$) goes from its first excited state to the ground state, it loses an amount of energy equal to

   (a) 10.2 eV

   (b) 13.6 eV

   (c) 40.8 eV

   (d) 54.4 eV

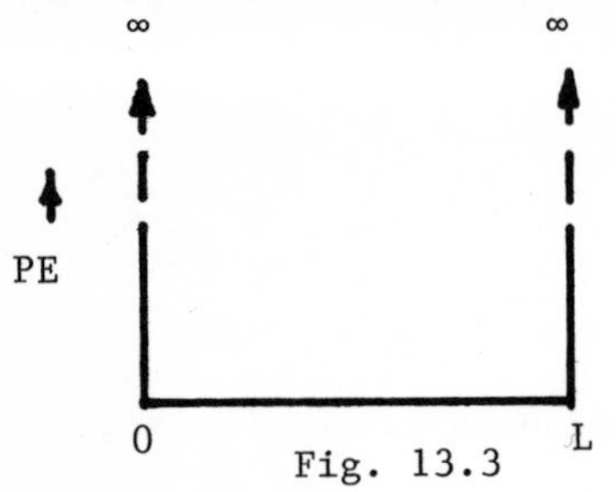

Fig. 13.3

4. An electron of mass m is confined to a one dimensional infinite well of length L. In Fig. 13.3 sketch the wave function for n = 3.

5. In Problem 4, when the electron is in the n = 3 state, its wavelength is

(a) 2L/3

(b) L

(c) 2L

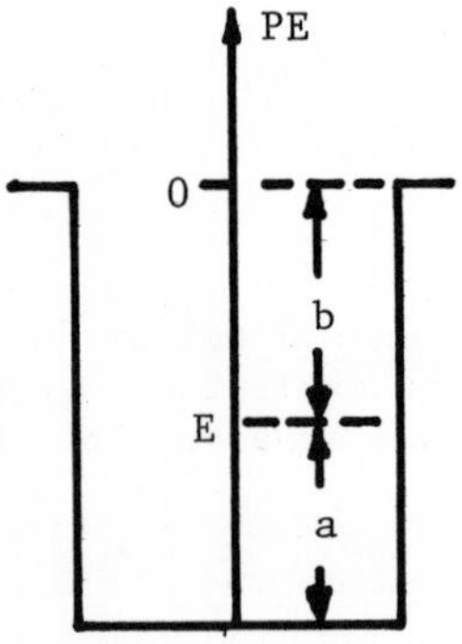

Fig. 13.4

6. An electron of mass m is confined to a one-dimensional well with finite walls (Fig. 13.4). It has a total energy E. In terms of the labels used in the figure, its kinetic energy is

(a) a

(b) b

(c) (a + b)

(d) (b - a)

(e) none of the above

7. The outer electron of lithium has an ionization energy of

$$E = 13.6\ Z^2_{eff}/n^2$$

The value of n in this formula for lithium is

(a) 1

(b) 2

(c) 3

(d) 4

8. The measured value for the ionization energy of lithium is 4.5 eV. The value of Z_{eff} for the outer electron is

(a) 0.58

(b) 1.15

(c) 1.73

(d) 3.0

9. In the ground state of palladium, all electronic states with $n \lesssim 4$ and $\ell \lesssim 2$ are occupied. The atomic number of palladium is

(a) 26

(b) 30

(c) 36

(d) 46

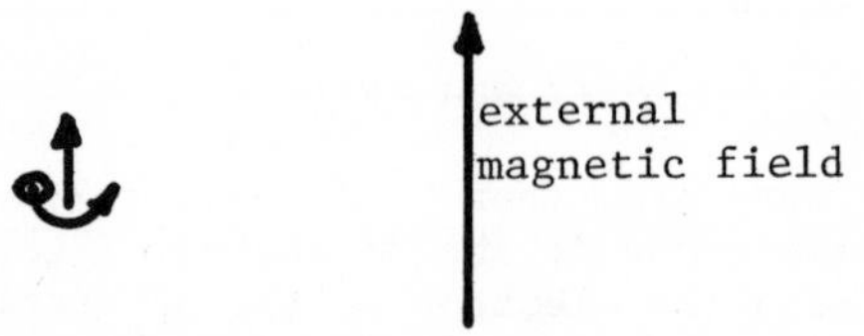

Fig. 13.5

10. The spin angular momentum of an electron is in the same direction as an external magnetic field (Fig. 13.5). The potential energy of this electron is

(a) less than an electron whose spin angular momentum is opposite to the external magnetic field

(b) equal to that of an electron whose spin angular momentum is opposite to the external magnetic field

(c) greater than an electron whose spin angular momentum is opposite to the external magnetic field

11. Imagine a world in which m_s could take on values

$$-5/2, -3/2, +1/2, +3/2$$

In such a world the valence electron in potassium (Z = 19) would be in which of the following states

(a) 2p

(b) 3s

(c) 3p

(d) 3d

(e) 4s

12. The volume available for an electron in a spherical shell depends on

(a) $1/r$

(b) r

(c) r^2

(d) r^3

where r is the distance from the nucleus.

13. The function ψ^2 is

(a) the probability of finding the electron

(b) the probability density

CHAPTER 14. The Nucleus

Objectives

You should learn

1. that the mass of a nucleus is slightly less than the sum of the rest masses of its constituent protons and neutrons
2. the sign of the attractive nuclear force, the repulsive electric force, and the exclusion principle contributions to nuclear energy
3. how to calculate the binding energy per nucleon for a given nucleus
4. how to calculate the disintegration energy Q for a nuclear decay
5. the evidence for the existence of the neutrino
6. how to write an equation for a nuclear reaction
7. the meaning of the decay constant λ and half life $T_{1/2}$.
8. when you can ignore the kinetic energy of the daughter nucleus
9. methods of detecting the age of organic compounds and rocks
10. the difference between fission and fussion

Trouble Spots

In finding the binding energy of nuclei, you may use atomic masses. It is true that you ignore the binding energy of the electrons, but these are small compared to nuclear binding energies. Atomic binding energies are of the order of magnitude of 10 or 20 eV, while nuclear binding energies are several MeV.

Possible decay schemes for the natural radioactive substances are:

alpha decay $\quad {}^{A}_{Z}X \rightarrow {}^{A-4}_{Z-2}Y + {}^{4}_{2}He$

Beta minus $\quad {}^{A}_{Z}X \rightarrow {}^{A}_{Z+1}Y + e^{-} + \bar{\nu}$

Beta plus $\quad {}^{A}_{Z}X \rightarrow {}^{A}_{Z-1}Y + e^{+} + \nu$

Electron capture $\quad {}^{A}_{Z}X + e^{-} \rightarrow {}^{A}_{Z+1}Y + \nu$

In the above equations, X is the symbol for the parent nucleus, Y the symbol for the daughter nucleus, e^- the symbol for an electron, e^+ for a positron, ν for a neutrino and $\bar{\nu}$ for an antineutrino.

In all nuclear reactions, check for conservation of charge and number of nucleons. For example, in alpha decay, the number of protons before is Z and the number after is

$$(Z - 2) + 2 = Z$$

The number of nucleons before is A and the number of nucleons after is

$$(A - 4) + 4 = A$$

A nucleon can be either a proton or a neutron.

A neutrino is involved in beta plus decay and electron capture. An antineutrino is associated with beta minus decay.

For beta minus and beta plus decay, the electron or positron has maximum energy E_{max} when the neutrino energy $E_\nu = 0$, when the energy of the electron or positron E_e is less than E_{max}, one neutrino accounts for the missing energy, momentum ($p_\nu = E_\nu/c$) and angular momentum ($h/2\pi$).

Unlike the other reactions, the electron mass m_e does not cancel out for beta plus decay. For beta plus decay,

$$Q = (M_X - Zm_e)c^2 - (M_Y - (Z - 1)m_e)c^2 - m_ec^2$$

or

$$Q = (M_X - M_Y - 2m_e)c^2$$

where M_X and M_Y are the atomic masses of the parent and daughter, respectively, and m_e is the mass of the electron.

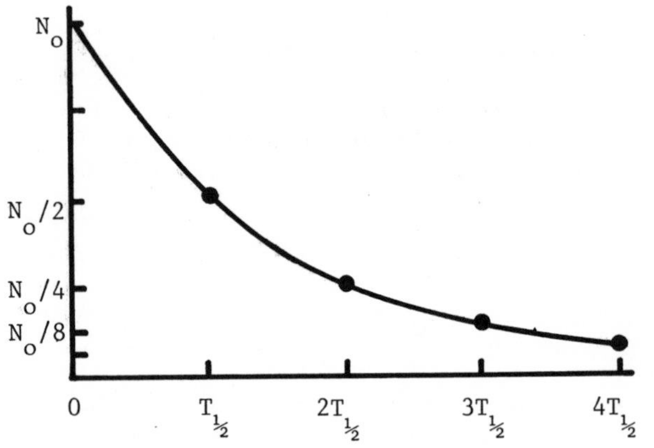

Fig. 14.1

If at time t = 0, N radioactive nuclei exist there will be N/2 remaining after one half life, N/4 after two half lives, N/8 after three half lives, and so forth. A plot of N as a function of time is shown in Fig. 14.1.

In alpha decay, the kinetic energies of the alphas may appear in discrete values. When this happens, the daughter nucleus is left in an excited state and there is subsequent emission of gamma rays.

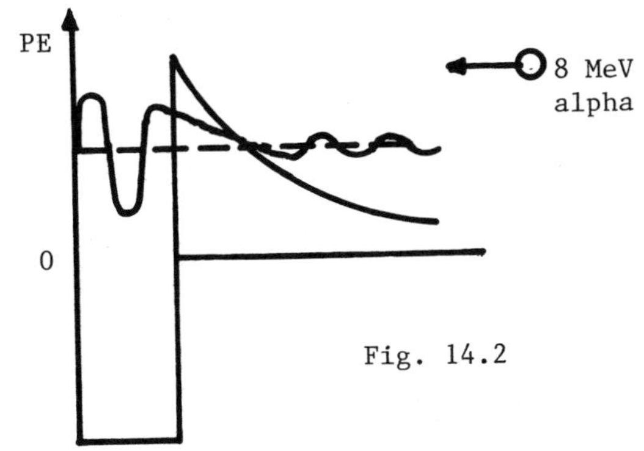

Fig. 14.2

While an alpha particle with an energy of 8 MeV may not be able to penetrate a certain nucleus, this nucleus may emit an alpha of approximately 6 MeV. This experimental fact may be understood in terms of the wave properties of particles. When the alpha particle is confined to the nucleus, its probability function does not drop to zero at the edges of the well. Although the probability is small there is a finite probability of finding the alpha particle outside the nucleus, as is shown in Fig. 14.2. The probability of emission increases with an increase in energy of the alpha particle. Thus, it is found that alpha emitters with high energy alpha's have a shorter half life.

When particle x bombards particle X producing particles y and Y, it might appear that the reaction can take place if the total energy of x (rest mass energy + kinetic energy) plus the rest mass energy of X is greater than the rest mass energy of (y + Y). This cannot happen because the momentum of the decay particles cannot equal zero. The momentum of the decay particles after the reaction must equal the momentum of x before the reaction.

<u>Illustrative Examples</u>

1. Given the following rest-mass energies of

the deuteron = 1875.62 MeV = M_D

the proton = 938.28 MeV = M_p

the neutron = 939.57 MeV = M_n

What is the binding energy of the deuterium nucleus?

The total binding energy is the sum of the masses of the individual nucleons minus the mass of the nucleus. Thus

$$BE = (M_p + M_n) - M_D$$

$$= (938.28 + 939.57)\text{MeV} - 1875.62 \text{ MeV}$$

$$= 2.23 \text{ MeV}$$

2. If the half life of radium is 1600 years, what fraction of a sample of radium will have decayed after 3200 years.

If N_0 radioactive nuclei are present at t = 0, after 1600 years (one half life),

$N_0/2$ remain.

After 3200 years (two half lives),

$N_0/4$ remain.

Thus the number of nuclei that have decayed is

$$N_0 - N_0/4 = 3/4\ N_0$$

and the fraction that has decayed is

$$3/4\ N_0/N_0 = 3/4$$

3. A common thermonuclear reaction is

$${}^1_0n + {}^6_3Li \rightarrow {}^3_1H + X$$

What is X?

Before the reaction, the number of protons is

$$0 + 3 = 3$$

After the reaction, the number Z of protons for X must be such that

$$1 + Z = 3$$

or

$$Z = 2$$

Before the reaction, the number of nucleons is

$$1 + 6 = 7$$

After the reaction, the number A of nucleons for X must be such that

$$3 + A = 7$$

or $$A = 4$$

Thus, X has 2 protons and 4 nucleons and is identified as helium. The reaction is

$${}^{1}_{0}n + {}^{6}_{3}Li \rightarrow {}^{3}_{1}H + {}^{4}_{2}He$$

4. ${}^{48}_{23}V$ decays by beta plus and electron capture. Find the Q of the reaction for the two types of decay. Given the following masses in atomic mass units

mass of ${}^{48}_{23}V$ = 47.947591

mass of ${}^{48}_{22}T$ = 47.952259

mass of the electron = 0.000549

For electron capture, Q = mass of ${}^{48}_{23}V$ minus the mass of ${}^{48}_{22}T$.

mass of ${}^{48}_{23}V$ = 47.947591

mass of ${}^{48}_{22}T$ = 47.952259

mass of ${}^{48}_{23}V$ - mass of ${}^{48}_{22}T$ = 0.004668 amu

Thus, for electron capture,

$$Q = 0.004668 \text{ amu} \times 931.5 \text{ MeV/amu}$$

$$= 4.35 \text{ MeV}$$

For beta plus, Q = mass of ${}^{48}_{23}V$ - mass of ${}^{48}_{22}T$ - two times the mass of the electron.

mass of ${}^{48}_{23}V$ - mass of ${}^{48}_{22}T$ = 0.004668

2 x mass of electron = 0.001098

Q = 0.003570 amu

= 3.33 MeV

5. A ${}^{8}_{4}Be$ nucleus of mass m approximately equal to $8 \times 1.67 \times 10^{-27}$ kg may decay to the ground state with the emission of a 17.6 MeV gamma ray. With what kinetic energy does the ${}^{8}_{4}Be$ nucleus decay?

The momentum of the gamma ray, with E_γ equal to the energy of the gamma ray, is

$$p = E_\gamma/c$$

$$= \frac{17.6 \times 10^{6} \text{ eV} \times 1.6 \times 10^{-19} \text{ J eV}^{-1}}{3 \times 10^{8} \text{ m s}^{-1}}$$

$$= 5.87 \times 10^{-2} \times 1.6 \times 10^{-19} \text{ J m}^{-1} \text{ s}$$

$$= 5.87 \times 10^{-2} \times 1.6 \times 10^{-19} \text{ kg-m s}^{-1}$$

By conservation of momentum, the nucleus must recoil with momentum $p = p_\gamma$.

$$p = mv$$

and

$$KE = 1/2 \; mv^2$$

$$= (mv)^2/2m$$

$$KE = \frac{(5.87 \times 10^{-2} \times 1.6 \times 10^{-19} \text{ kg-m s}^{-1})^2}{2 \times 8 \times 1.67 \times 10^{-27} \text{ kg}}$$

$$= \frac{(5.87 \times 10^{-2} \times 1.6 \times 10^{-19})^2 \text{ J}}{26.72 \times 10^{-27}}$$

$$= \frac{(5.87 \times 10^{-2} \times 1.6 \times 10^{-19})^2 \text{ J}}{26.7 \times 10^{-27} \times 1.6 \times 10^{-19} \text{ J-eV}^{-1}}$$

$$= \frac{(5.87 \times 10^{-2})^2 \times 1.6 \times 10^{-19} \text{ eV}}{26.72 \times 10^{-27}}$$

$$= 2.06 \times 10^4 \text{ eV}$$

Problems

1. The mass of the nucleus is

   (a) slightly less

   (b) equal to

   (c) greater than

   the sum of the rest masses of its constituent particles.

2. The density of a nucleus

   (a) decreases with atomic number

   (b) is practically independent of the atomic number

   (c) increases with atomic number

3. The decay constant

   (a) depends on how long the nucleus has existed

   (b) depends on the presence of other nuclei

   (c) is independent of the above two quantities

4. After 4 half lives the fraction of a sample that remains is

   (a) 1/4

   (b) 1/8

   (c) 1/16

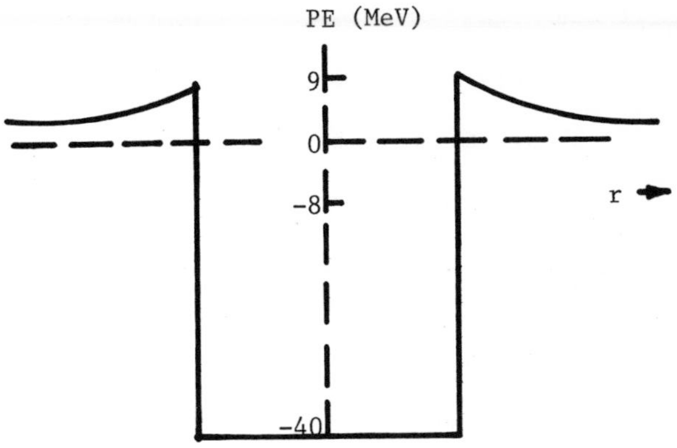

Fig. 14.3

5. Figure 14.3 is a sketch of a potential well for a copper nucleus. A proton p confined to the nucleus has a total energy E = -8 MeV. The kinetic energy of this proton is

   (a) 8 MeV

   (b) 9 MeV

   (c) 32 MeV

6. If the proton in Problem 5 absorbs a 15 MeV gamma ray, the copper nucleus can decay to ^{62}Ni plus a proton. The kinetic energy of the proton when it is far from the nucleus is

   (a) 2 MeV

   (b) 7 MeV

   (c) 15 MeV

7. In the reaction,

$$^{14}_{7}N + X \rightarrow {}^{14}_{6}C + {}^{1}_{1}H,$$

   X is ______________

8. If a nucleus can decay by electron capture, it can

(a) sometimes decay by beta minus

(b) always decay by beta minus

(c) never decay by beta minus

9. When a certain substance decays, it emits alpha particles of kinetic energy 5.31 MeV and 5.60 MeV. The difference in energy of the alphas is accounted for by

(a) a neutrino with energy 0.29 MeV

(b) a gamma ray with wavelength $\lambda = 2.34 \times 10^{-13}$ m

(c) a gamma ray with wavelength $\lambda = 42.76 \times 10^{-13}$ m

10. ^{237}Np is radioactive. The ultimate decay product is

(a) ^{208}Pb

(b) ^{209}Bi

(c) ^{206}Pb

11. Assume two deuterons must get as close as 10^{-14} m in order for the nuclear force to overcome the repulsive electrostatic force. What is the height in million electron volts of the electrostatic potential barrier?

(a) 0.14 MeV

(b) 0.14×10^{14} MeV

(c) 2.3×10^{-15} MeV

12. In Problem 11, the temperature to which the deuterium must be heated so that the deuterons may overcome the repulsive electrostatic force is

(a) 1.1×10^{9} degrees K

(b) 1.1×10^{23} degrees K

CHAPTER 15. Particle Physics

Objectives

You should learn

1. how to draw Feynman diagrams for

    (a) the electromagnetic interaction

    (b) the strong interaction

    (c) the weak interaction

2. how to use the uncertainty relation between the uncertainty in energy and the uncertainty in time

$$\Delta E \Delta t \gtrsim h/2\pi$$

to estimate the range of the four interactions

3. the relative strengths of the four interactions

4. which interaction is important in a process that involves a neutrino

5. the conservation laws for "additive quantum numbers":

    (a) electric charge Q

    (b) electron number N_e

    (c) muon number N_μ

    (d) baryon number B

    (e) strangeness S

    (f) hypercharge Y = S + B

6. that antiparticles have opposite values to the particles for all of the additive quantum numbers

7 the conservation laws "obeyed" by the four interactions

8. that any process that satisfies

$$\Delta S = \pm 1$$

can occur via the weak interaction, but cannot occur through the strong interaction

9. how to arrange super multiplets by hypercharge Y and isotopic spin component I_z

10. the properties of quarks

11. how to form baryons and mesons with combinations of quarks

12. the meaning of invariance properties

13. the meaning of the parity operation as a reflection in a mirror

14. the meaning of charge conjugation as the replacement of every particle by its antiparticle

15. the meaning of time reversal as the reversal of the direction of time

16. how to apply the above three operations

Trouble Spots

The conservation of energy can be violated within the limit of the time-energy uncertainty principle

$$\Delta E \Delta t \gtrsim h/2\pi$$

The time-energy uncertainty principle is involved in estimating the range of a force. Since the mass-energy of the exchange particle is usually given in MeV, it is useful to express $h/2\pi$ in Mev-s.

$$h/2\pi = \frac{6.63 \times 10^{-34}\ \text{J-s}}{2\pi(1.6 \times 10^{-13}\ \text{J-MeV}^{-1})}$$

$$= 6.58 \times 10^{-22}\ \text{MeV-s}$$

A virtual pion of rest mass energy 140 MeV can exist for a time

$$\frac{h}{2\pi(140\ \text{MeV})} = \frac{6.58 \times 10^{-22}\ \text{MeV-s}}{140\ \text{MeV}}$$

$$\Delta t = 4.7 \times 10^{-24}\ \text{s}$$

If we assume that the pion travels with a speed that is approximately equal to the speed of light, c, the range of the pion is

$$\text{Range} = \Delta t \times c$$

$$= 4.7 \times 10^{-24}\ \text{s} \times 3 \times 10^{8}\text{m s}^{-1}$$

$$= 1.4 \times 10^{-15}\ \text{m}$$

You may argue that we used the rest mass of the pion and then later said the pion travels with a speed that is approximately equal to the speed of light. Let us repeat the calculation with the speed of the pion equal to v = 0.995 c. Then the energy of the pion is

$$E = \frac{m_0 c^2}{\sqrt{1 - (0.995)^2}}$$

$$= \frac{140\ \text{MeV}}{\sqrt{1 - 0.99}}$$

$$= \frac{140\ \text{MeV}}{\sqrt{0.01}}$$

$$= \frac{140\ \text{MeV}}{0.1}$$

$$= 1400\ \text{MeV}$$

Now, $$\Delta t = \frac{6.58 \times 10^{-22}\ \text{MeV-s}}{1400\ \text{MeV}}$$

$$= 1.4 \times 10^{-16}\ \text{s}$$

and range $= \Delta t \times 0.995 \times c$

$$= 4.7 \times 10^{-25}\ \text{s} \times 0.995 \times c$$

$$= 4.7 \times 10^{-25}\ \text{s} \times 2.865 \times 10^{8}\ \text{m s}^{-1}$$

$$= 1.35 \times 10^{-16} \text{ m}$$

Since in using the uncertainty relation, we are already using an approximation, we may ignore the difference between 1.4×10^{-15} m and 1.35×10^{-16} m.

In "associated production", processes that proceed by the strong interaction, strangeness S (and hypercharge Y) is conserved. A possible production process in a bubble chamber is

$$\pi^- + p \rightarrow K^+ + \Sigma^-$$

S: $0 + 0 = +1 + (-1)$

Y: $0 + 1 = +1 + 0$

The decay processes via the weak interaction do not conserve strangeness and hypercharge.

$$\Sigma^- \rightarrow \pi^- + n^o$$

S: $-1 \neq 0 + 0$

Y: $0 \neq 0 + 1$

and

$$K^+ \rightarrow \pi^+ + \pi^o$$

S: $+1 \neq 0 + 0$

Y: $+1 \neq 0 + 0$

Any process that satisfies $\Delta S = \Delta Y = \pm 1$ can take place via the weak interaction. Some examples are:

$$K^- \rightarrow \pi^- + \pi^o$$

S: $-1 \neq 0 + 0$

Y: $-1 \neq 0 + 0$

$$\Lambda^o \rightarrow \pi^- + p$$

S: $-1 \neq 0 + 0$

Y: $0 \neq 0 + 1$

Processes with $\Delta S = \pm 2$ cannot occur. For example the following process does not occur:

$$\Xi^- \rightarrow n + \pi^-$$

S: $-2 \neq 0 + 0$

Y: $-1 \neq 0 + 1$

Baryon number, electron number, and muon number are conserved in all decays, as is shown in the following examples:

$$n \rightarrow p + e^- + \bar{\nu}_e$$

B: $1 = 1 + 0 + 0$

N_e: $0 = -1 + 0 + (+1)$

N_μ: $0 = 0 + 0 + 0$

$$\mu^+ \rightarrow e^+ + \bar{\nu}_\mu + \nu_e$$

B: $0 = 0 + 0 + 0$

N_e: $0 = -1 + 0 + 1$

N_μ: $-1 = 0 + (-1) + 0$

$$\Sigma^- \rightarrow n + \pi^-$$

B: $1 = 1 + 0$

N_e: $0 = 0 + 0$

N_μ: $0 = 0 + 0$

The following strong interaction fails to occur only because the kaon is too massive.

$$\Omega^- \not\rightarrow \Xi^o + K^-$$

Q: $-1e = 0 + (-1e)$

B: $+1 = +1 + 0$

S: $-3 = -2 - 1$

Y: $-2 = -1 - 1$

Rest Mass: $1673 \text{ MeV} < (1321 + 494) \text{MeV}$

The component of isospin

$$I_z = Q/e - Y/2$$

$$= Q/e - (B + S)/2$$

where Q equals the charge of a particle in multiples of e, B the baryon number, Y the hypercharge, and S the strangeness. Since Y = B + S, the conservation of baryon number and strangeness implies the conservation of hypercharge. The conservation of charge and strangeness implies the conservation of the component of isotopic spin. Isotopic spin is useful in determining the number of members of a multiplet. For example, we know that

Σ^- exists with B = 1, S = -1 and Q = -e.

For the Σ^-,

$$I_z = -e/e - (1 - 1)/2$$

$$= -1$$

Isospin multiples have I_z from

$$-I, (-I + 1), \ldots 0, 1, 2, \ldots I$$

For $I_z = -1$, there must exist other particles with essentially the same rest mass with I_z = 0 and $I_z = 1$. These correspond to $\Sigma^0(I_z = 0)$ and $\Sigma^+(I_z = +1)$.

The invariance principle means that given an experiment that obeys the laws of physics, you can create other experiments that obey the same laws by carrying out a symmetry operation. In deciding whether an operation is valid, you should check to see if the results of the experiment can exist in "our world".

Illustrative Examples

1. Another possible exchange particle for the strong interaction is the kaon of rest mass energy equal to 494 MeV. What is the range of the nuclear force for the kaon?

   In general,

$$\Delta t \Delta E = h/2\pi = 6.58 \times 10^{-22} \text{ MeV-s}$$

For the kaon with rest mass energy = 494 MeV,

$$t = \frac{6.58 \times 10^{-22} \text{ MeV-s}}{494 \text{ MeV}}$$

$$= 1.33 \times 10^{-24} \text{ s}$$

Assuming the kaon moves at a speed near the speed of light, we find for the range

$$r = \Delta t\, c$$

$$= 1.33 \times 10^{-24} \text{ s} \times 3 \times 10^8 \text{ m s}^{-1}$$

$$= 4 \times 10^{-16} \text{ m}$$

2. Write down the antiparticle for

   (a) π^+, (b) π^0, (c) K^+ (d) K^0 (e) ν_e (f) Σ^+

   (a) The antiparticle of π^+ is the π^-

   (b) The antiparticle of the π^0 is itself

   (c) The antiparticle of the K^+ is the K^-

   (d) The antiparticle of the K^0 is the $\bar{K}^0$

   (e) The antiparticle of the ν_e is the $\bar{\nu}_e$

   (f) The antiparticle of the Σ^+ is the $\bar{\Sigma}^-$

3. A proton is incident upon a stationary proton in a bubble chamber. What is the total rest energy that can be produced?

We assume that a single intermediate particle is produced. The single intermediate particle may not really exist or it may and then decay into two or more particles. The rest mass m_0' of the particle is related to its total energy E' and momentum p' by

$$m_0' c^2 = \sqrt{E'^2 - (p'c)^2}$$

If E_p is the total energy of the incident particle and m_0 is the rest mass of the proton, conservation of energy gives

$$E' = E_p + m_0c^2$$

And conservation of momentum gives

$$pc' = p_pc$$
$$= \sqrt{E_p^2 - (m_0c^2)^2}$$

Thus $m_0'c^2 = \sqrt{E'^2 - (p'c)^2}$

can be written as

$$m_0'c^2 = \sqrt{(E_p + m_0c^2)^2 - (E_p^2 - m_0c^2)^2}$$

or

$$(m_0'c^2)^2 = E_p^2 + 2E_pm_0c^2 + m_0^2c^4 - (E_p^2 - m_0^2c^4)$$
$$= 2m_0c^2\ (E_p + m_0c^2)$$
$$= 2m_0c^2\ (KE_p + 2m_0c^2)$$

where the kinetic energy of the proton KE_p equals $(E_p - m_0c^2)$, where E_p is the total energy of the incident proton and m_0 is its rest mass. For example, if the total energy of the incident proton $E_p = 449\ m_0c^2$, then

$$(m_0'c^2)^2 = 2m_0c^2(449\ m_0c^2 + m_0c^2)$$
$$= 900\ (m_0c^2)^2$$

and $(m_0'c^2) = 300\ m_0c^2$

Notice that the total energy available for the rest mass of the particle is less than $E_p + m_0c^2$ because some of the energy must go into energy of motion to conserve momentum.

4. Classify the following reactions as to the type of interaction taking place:

(a) $\mu^+ \rightarrow e^+ + \nu_e + \bar{\nu}_\mu$

(b) $e^- + p \rightarrow \gamma + {}^1_1H$

(c) $p + \bar{p} \rightarrow \Lambda + \bar{\Lambda}$

(a) This must be a weak interaction because it involves a neutrino.

(b) The signal here is the gamma ray. It is an electromagnetic interaction.

(c) For the reaction,

$$p + \bar{p} \rightarrow \Lambda + \bar{\Lambda}$$

Q:	1e + (−1e)	=	0 + 0
B:	1 + (−1)	=	1 + (−1)
S:	0 + 0	=	(−1) + 1

All of the additive quantum numbers are conserved in this reaction. It is a strong interaction.

5. The following decay modes are forbidden because they violate one of the conservation laws. For each decay, indicate which conservation law is violated.

(a) $p \rightarrow e^+ + \nu_e$

(b) $n \rightarrow e^+ + p + \nu_e$

(c) $n \rightarrow e^- + p + \nu_e$

(d) $p \rightarrow n + e^+ + \nu_e$

(e) $\Xi^- \rightarrow \pi^- + n$

In the following, we write the charge in units of e. That is, if a particle has charge +1e, we write it as +1. If it has charge −1e, we write it as −1.

(a) $p \rightarrow e^+ + \nu_e$

Q	1	=	1	+	0
B	1	≠	0	+	0
N_e	0	=	−1	+	1
S	0	=	0	+	0

In this reaction, conservation of baryon number is violated.

(b) $n \rightarrow e^+ + p + \nu_e$

Q	$0 \neq 1 + 1 + 0$
B	$1 = 0 + 1 + 0$
N_e	$0 = -1 + 0 + 1$
S	$0 = 0 + 0 + 0$

In this reaction, conservation of charge is violated.

(c) $n \rightarrow e^- + p + \nu_e$

Q	$0 = -1 + 1 + 0$
B	$1 = 0 + 1 + 0$
S	$0 = 0 + 0 + 0$
N_e	$0 \neq 1 + 0 + 1$

In this reaction, conservation of electron number is violated

(d) $p \rightarrow n + e^+ + \nu_e$

Q	$1 = 0 + 1 + 0$
B	$1 = 1 + 0 + 0$
N_e	$0 = 0 - 1 + 1$
S	$0 = 0 + 0 + 0$

Although it appears, that this reaction obeys all of the conservation laws, it cannot happen because in terms of the rest mass energies

938.3 MeV < 939.6 MeV + 0.51 MeV

(e) $\Xi^- \rightarrow \pi^- + n$

Q	$-1 = -1 + 0$
B	$1 = 0 + 1$
N_e	$0 = 0 + 0$
S	$-2 \neq 0 + 0$

This reaction cannot take place because even for a weak interaction, $\Delta S = \pm 1$

6. Use the following table to decide how to build a proton from three quarks

Quark	Q	B	Y	I_z
u	+2e/3	+1/3	+1/3	1/2
d	-e/3	+1/3	+1/3	-1/2
s	-e/3	+1/3	-2/3	0

For the proton p

	Q	B	Y	I_z
p	e	1	1	1/2

Notice for 2 u quarks, the quantum numbers are

Q	B	Y	I_z
+4e/3	+2/3	+2/3	1

If we add the quantum numbers for an s quark, for the combination of 2u + s, the quantum numbers are

	Q	B	Y	I_z
2u + s	e	1	1	1/2

Note the quantum numbers for 2u + s are identical to those for a proton.

7. A K^+ decays to a μ^+ and ν_μ with a half life of 1.2×10^{-8} s. (a) What is the minimum uncertainty in the rest mass energy? (b) Draw the linear and spin angular momentum vectors assuming the K^+ is initially at rest. The spin angular momentum of the kaon is zero. (c) Draw the mirror image of this decay and state whether the mirror image is a possible decay. (d) Draw the charge conjugation reaction of the mirror image decay and state whether it is a possible decay.

(a) $\Delta E \Delta t = h/2\pi$

$$= 6.58 \times 10^{-22} \text{ MeV-s}$$

and $\Delta E = 6.58 \times 10^{-22}$ MeV-s/ Δt

$$= \frac{6.58 \times 10^{-22} \text{ MeV-s}}{1.2 \times 10^{-8} \text{ s}}$$

$$= 5.48 \times 10^{-14} \text{ MeV}$$

(b) The decay is shown in Fig. 15.1. Since the muon neutrino is left handed, point the thumb of the left hand in the direction of its motion and your fingers give the direction of the spin. Since the angular momentum of the kaon is zero, the positive muon must spin in the opposite direction to keep the angular momentum after the decay equal to zero.

Fig. 15.1

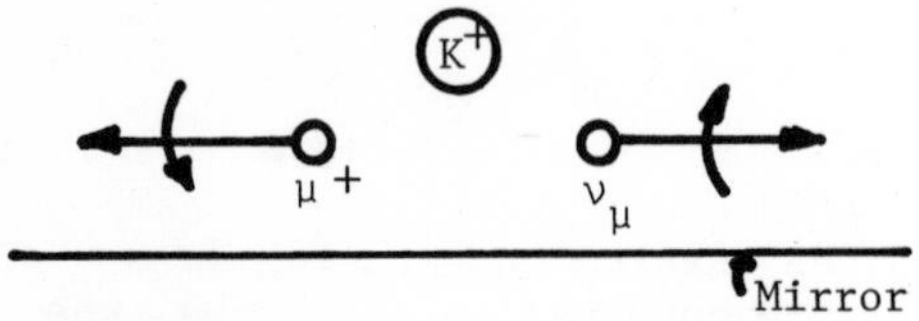

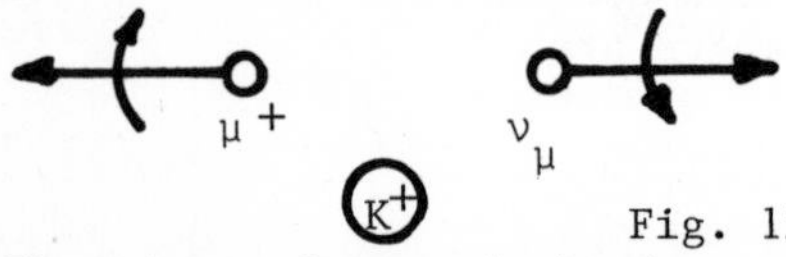

Fig. 15.2

(c) The mirror image of the decay is shown in Fig. 15.2. In the mirror image of the neutrino you must point the thumb of your right hand in the direction of the neutrino to get the mirror reflected direction of spin. Since the neutrino is left handed, the decay of the plus kaon to a plus muon and a muon neutrino is not invariant under mirror reflection.

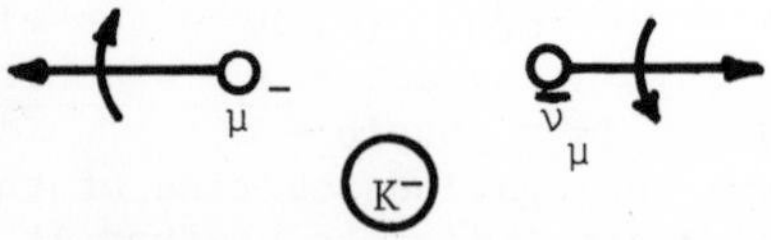

Fig. 15.3

(d) To charge conjugate the reaction, change all the particles to their antiparticles. The reaction is then

$$K^- \rightarrow \mu^- + \bar{\nu}_\mu$$

The charge conjugation of the mirror image is shown in Fig. 15.3. Now when you point the thumb of your right hand in the direction of the motion of the antineutrino your fingers give the direction of the spin. Since the antineutrino is right handed CP invariance is allowed. While reflection in a mirror, P operation, is not possible, and charge conjugation, C operation is not possible, the combined symmetry operations of CP are possible.

Problems

1. The rest mass energy of π^0 is 135 MeV. The rest mass energy of η^0 is 549 MeV. The contribution of pions to the nuclear force is

   (a) less than etas

   (b) equal to that of the etas

   (c) greater than etas

2. Which of the following particles are stable?

   (a) neutron

   (b) muon

   (c) electron

3. The interaction that violates the most conservation laws is

   (a) the strong

   (b) electromagnetic

   (c) weak

   (d) gravitational

4. The following strong interaction takes place

$$K^- + p \rightarrow K^+ + X$$

The identity of X is

(a) Σ^-

(b) Ξ^-

(c) Ω^-

5. The mass of the resonance called (1530 MeV) has a range of uncertainty of 7.14 MeV. At a speed of 0.1 c, the distance traveled by this particle before it decays is

(a) 1.29×10^{-17} m

(b) 3.0×10^{-16} m

(c) 2.76×10^{-15} m

6. Imagine that particle X^{++} with a rest mass energy of 2000 MeV is produced when a positive pion (rest mass 140 MeV) collides with a stationary proton. Take the rest mass of the proton to be 1000 MeV. The minimum kinetic energy of the pion to produce X^{++} is

(a) 600 MeV

(b) 1350 MeV

(c) 2000 MeV

7. The measurements of the rest mass of the X^{++} particle in Problem 6 varies over a range of 60 MeV. The interaction that causes its decay is

(a) strong

(b) electromagnetic

(c) weak

8. The following decay is forbidden.

$$\Lambda^0 \rightarrow p + \pi^-$$

The conservation law that is violated is

(a) charge

(b) baryon number

(c) strangeness

(d) energy

9. Particle X^0 is formed in the reaction

$$K^- + p \rightarrow X^0 + K^0 + K^0$$

There must exist

(a) no other states of X

(b) one other state of X

(c) two other states of X

10. The K^- has spin 0 and can decay by the reaction

$$K^- \rightarrow \mu^- + \bar{\nu}_\mu$$

The decay is invariant under the operation

(a) P

(b) C

(c) PC

11. Suppose that three mesons, the V^0, V^-, V^{--}, all with nearly the same mass, have been discovered. The hypercharge of these mesons is

(a) 0

(b) -1

(c) 1

(d) -2

(e) 2

Sample Quiz for Chapters 13 through 15

1. In the ground state of the Bohr model of the hydrogen atom, the potential energy is

(a) -13.6 eV

(b) +13.6 eV

(c) -27.2 eV

2. The wavelength of the K_α line for an element is 0.7×10^{-10} m. The element is

(a) copper

(b) molybdenum

(c) tungsten

3. The energy of the lowest level for a one-dimensional well with infinite sides is E_1. The energy of the third level is

(a) $E_1/9$

(b) $3E_1$

(c) $9E_1$

4. Which of the following is true?

(a) the 3s level of hydrogen is lower than the 3s level of sodium

(b) the 3s level of hydrogen is the same as the 3s level of sodium

(c) the 3s level of hydrogen is higher than the 3s level of sodium

5. The ultimate decay product of ^{238}U is

(a) ^{208}Pb

(b) ^{206}Pb

(c) ^{207}Pb

6. The half life of carbon 14 is about 6000 years. The activity of the carbon in a certain biological relic is one-fourth that of a recent specimen. The age of the relic is

(a) 6000 years

(b) 12,000 years

(c) 18,000 years

(d) 24,000 years

7. The maximum kinetic energy of the neutrino is

(a) greater for beta plus decay than it is for electron capture

(b) the same for beta plus decay as it is for electron capture

(c) less for beta plus decay than it is for electron capture

8. In associated production the following reaction occurs

$$p + p \rightarrow X + K^o + p + \pi^+$$

X is

(a) $\bar{K}^o$

(b) Σ^-

(c) Λ^o

9. The following decay process takes place

$$\Xi^o \rightarrow \Lambda^o + \pi^o$$

The approximate life time of the Ξ^o is

(a) 10^{-8} s

(b) 10^{-16} s

(c) 10^{-23} s

10. The π^+ (spin = 0) decays by the following reaction

$$\pi^+ \rightarrow \mu^+ + \nu_\mu$$

The decay is invariant under the operation

(a) P

(c) C

(d) PC

Answers to Problems

Introduction. A Review of Mathematics

1. 2.5×10^{-5}
2. a + 2b
3. $x^2 - a^2$
4. 10^{-3}
5. 9×10^{16}
6. 2
7. 10^3
8. 2
9. $\frac{AB}{A + B}$
10. $\frac{A + B}{AB}$
11. 3/2
12. 81
13. 5 and -1
14. 7 and -2
15. x = 1/2, y = 2
16. x = 3, y = 4
17. 7/24
18. 24/25
19. -7/25
20. $^3\pi/8$
21. 1/2
22. $1/2kx^2$
23. $90\pi g$
24. $4\pi cm^2$

25. 0.5 cm

Chapter 1.

1. (b) An object may remain at rest even if individual forces act on the object as long as the net force equals zero.

2. (b) If your answer was (a) or (c), you are being an Aristotelian.

3. (b) The earth attracts two books with a greater force than one book. To keep the books at rest, you must exert a greater force to support two books.

4. (c) The ball continues forward, as it falls down, with the horizontal speed of the train.

5. (b) The force of friction.

6. (a) Aristotle believed that it took a net force to keep an object in motion.

7. (b) If this were not true, the theory of relativity would be violated.

8. (b) It takes a net force to change the direction of an object.

9. (b) Hay continues with horizontal velocity of the plane.

10. (b)

11. (a)

Chapter 2.

1. (a) If your answer was (b), you are confusing velocity and acceleration.

2. (b) $s = 1/2\ at^2 = 1/2 \times 10\ m\ s^{-2} \times 16\ s^2$

$= 80m.\ \bar{v} = s/t = 20\ m\ s^{-1}$

3. (c) $v = v_0 + at = 16\ m\ s^{-1}$

4. (a) $v^2 = 2\ as;\quad s = v^2/2a = 2\ m$

5. (c) Distance between positions decreases with time.

6. (b) This is a vector addition.

7.

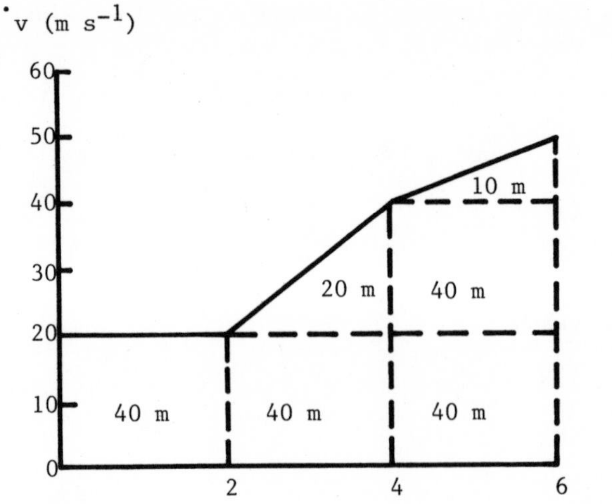

Fig. 2.6(b)

8. (c)

9. (b) $(v^2)_h = (v_0^2)_h + 2\ a_h\ s_h$

$0 = (v_0^2)_h + 2\ a_h\ s_h$

where $(v_0)_h = 10\ m\ s^{-1} \times \sin 53^o$

and $a_h = -10\ m\ s^{-2}$

10. (b)

Chapter 3

1. (c) a = net force/m

2. (c)

3. (a) The slope of the average velocity versus time is acceleration.

4. (b) 30 N - frictional force = m a

5. (c) The frictional force = 10 N since this is the force the man must exert to move the object with a constant velocity.

6. (c) $v^2 = 2$ (Net force/m) s

7. (a) The vector addition for force gives

a net force of 13 N, 22.5° north of east. Then a = net force/m

8. (c) Velocity toward east equals 4 m s^{-1}. t = s/v

9. (b) $v_N = (v_o)_N + a_N t$

10. (b)

Chapter 4

1. (c)

2. (b)

3. (c)

4. (c) Since there is an acceleration downward, the gravitational force must be greater than the force of the scales on the person.

5. (c) You must include the component of the weight of the block down the plane in finding the net force.

6. (b)

7. (c) In this case you cannot consider the gravitational force to be independent of the distance r.

8. (c) The net force acting on m is $F_1 - F_2$

9. (c) The period is the time for one complete vibration.

10. (c) At t = 0.5 s, the displacement is negative, so the acceleration is positive.

Chapter 5

1. (c) F = m g

2. (c) Because of the equivalence of the inertial and gravitational mass, an acceleration is identical to a gravitational force.

3. (b)

4. (b)

5. (c)

6. (a)

7. (b) The force of A on B is 5 N toward A. The force of C on B is 12 N toward C.

8. (a)

9. (e)

10. (b)

Chapter 6

2. (c) Field due to both charges is to the right.

2. (c)

3. (a) The electric force is independent of the velocity of the charge.

4. (b) The electric force does not change the horizontal velocity.

5. (c) Find the vertical acceleration and then use the time to find the vertical velocity.

6. (d) The magnetic force must be up, but the magnetic field must be into the page.

7. (c) The magnetic field due to both wires is down.

8. (b) r = mv/qB; r' = 8 mv/4qB

9. (c) The magnetic force is always perpendicular to the plane that contains the velocity and magnetic field.

10. (a) There is no magnetic force on a particle that moves parallel or antiparallel to the magnetic field.

Chapter 7

1. (b) Note that KE = $p^2/2m$

2. (a) Remember that momentum is a vector quantity.

3. (b) After the collision, the momentum of M to the east is 15 kg-m s^{-1} and 8 kg-m s^{-1} to the north. Since the initial momentum was 15 kg-m s^{-1} to the east, the wooden disk must travel south.

4. (b) Change in kinetic energy = change in potential energy = eV.

5. (c) Use the fact that the centripetal acceleration is produced by the electric force to find mv^2. Then E = KE + PE = KE $-(k_e e^2/r)$

6. (b) The total energy = E = 1/2 k A^2, where A is the amplitude. Then E = 1/2 kx^2 + 1/2 mv^2.

7. (b) You do not have to add atmospheric pressure because the pressure inside the sub is atmospheric pressure.

8. (c) The Kelvin temperature is proportional to the square of the average velocity.

9. (a)

10. (e) Angular momentum is conserved and $v = 2\pi r f$.

11. (e)

Chapter 8

1. (b)

2. (b) Don't forget to invert the reciprocal to find R.

3. (b) The current in R equals the electromotive force divided by the total resistance of the circuit.

4. (b)

5. (a) The current i equals ($\xi_1 - \xi_2$) divided by the total resistance of the circuit.

6. (b) The potential difference from a to c equals ($\xi_2 + ir_2$).

7. (b) 1.5 V = 10^{-3} A (500 + R) ohm

8.

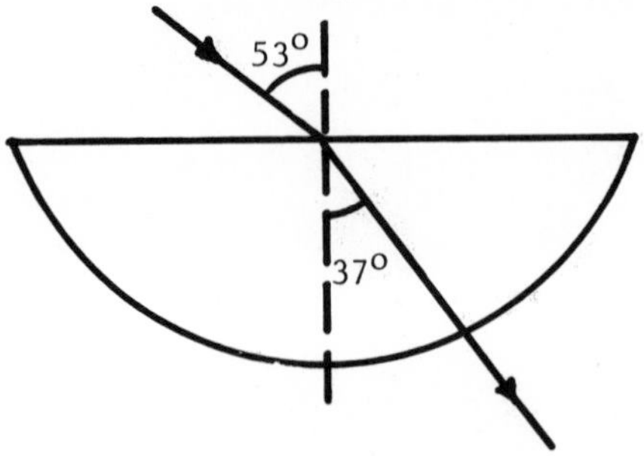

9. (c)

10. (a)

11. (a) m = q/p

12. (a) The position of the first image is 120 cm to the right of the converging lens or 100 cm to the right of the diverging lens. Since this is a virtual object take the object distance for the diverging lens equal to -100 cm.

Chapter 9

1. (c) In Fig. 9.7 the wave has traveled a distance of $\lambda/2$. Thus, the time in this figure corresponds to T/2.

2. (b) $v = \lambda/T$

3. (c) This is a traveling wave with $y = A \sin (2\pi s/\lambda - 2\pi t/T)$

4.

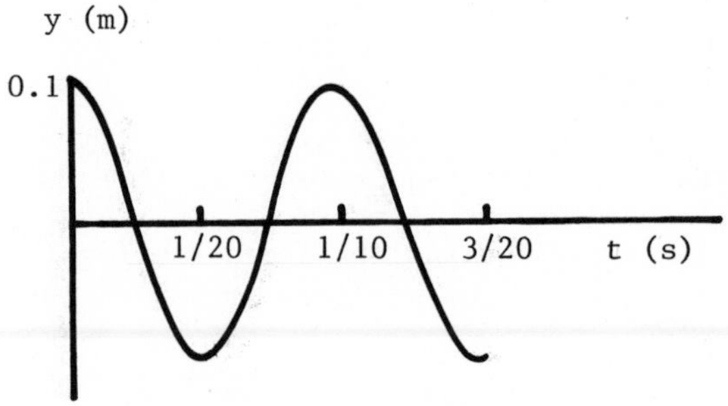

5. (d) This is a standing wave. In this case, $T/4$ has elapsed so $T = 1/5$ s. For a standing wave $y = A(\sin 2\pi s/\lambda)\cos(2\pi t/T)$

6. (b) Path difference $= \lambda/2$. $f = v/\lambda$

7. (c) $\Delta D = \lambda L/d$

8. (c) For a minimum, $m\lambda/N = w \sin\theta$, where m is an integer that is not a multiple of N.

9. (b)

10. (b)

11. (d) The resolving power is inversely proportional to wave length.

Chapter 10

1. (a)

2. (a)

3. (c) An increase in the speed of q produces an increase in the field due to the moving charge out of the page. This field opposes the increasing external field into the page.

4. (b)

5. (a) $i = \xi/R = B\ell v/R$

6. (b) $F = Bi\ell$

7. (c)

8. (d)

9. (d)

10. (a) $p = E/c$

Chapter 11

1. (a) The length of the spaceship for observers on the space platform is an improper length.

2. (a) This is a proper time.

3. (c) This is an improper time.

4. (c) This is a proper length $\Delta x' = \Delta x/\sqrt{1 - v^2/c^2}$

5. (c) $t' = \Delta x' \, v/c^2$

6. (a) $\Delta t' = \Delta t\sqrt{1 - v^2/c^2}$

7. (c)

8. (b) The velocities are not additive.

9. (a) $W = \Delta E$

$$= (m_o c^2/\sqrt{1 - v^2/c^2} - m_o c^2)$$

10. (c) Note that $M \neq 2\,m_o$

11. (a)

12. (b) $v/c = pc/E$

Chapter 12

1. (a) $E = hf$

2. (a) $p = hf/c$

3. (b) Some of the energy of the photon goes into removing the electron from the surface.

4. (c) An increase in intensity corresponds to an increase in the number of photons. More photons knock out more electrons.

5. (c) A loss in energy corresponds to an increase in wavelength.

6. (b)

7. (c) $hf = 1/3\, m_o c^2$; $f = m_o c^2/3h$

$$\lambda = c/f = 3h/m_o c$$

$$\lambda' = \lambda + 2h/m_o c = 5h/m_o c$$

$$f' = c/\lambda' = m_o c/5h$$

8. (b) $hf - hf' = m_0c^2 (1/\sqrt{1 - v^2/c^2} - 1)$

9. (c)

10. (c) $A = \vec{A}_1 + \vec{A}_2$

11. (a) In both cases $p = h/\lambda$

12. (a)

Chapter 13

1. (b) $-3.4eV - (-13.6eV) = 10.2eV$
$10.5eV - 10.2eV = 0.3eV$

2. (f) n = 4 to n = 1, n = 4 to n = 2, n = 4 to n = 3, n = 3 to n = 2, n = 3 to n = 1, n = 2 to n = 1

3. (c) For helium, $E_n = -13.6(4)eV/n^2$

4.

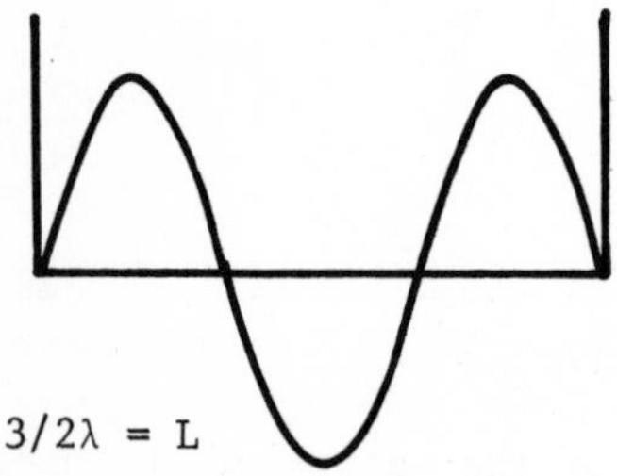

5. (a) $3/2\lambda = L$

6. (a) KE = E - PE = -b -(-a -b)

7. (b) Valence electron is in n = 2.

8. (b) $4.5\ eV = 13.6\ Z^2_{eff}/4$

9. (d) $1s^2 2s^2 2p^6 3s^2 3p^6 3d^{10}\ 4s^2 4p^6 4d^{10}$

10. (c) For this case the spinning electron acts like a north-seeking pole antialigned with the external magnetic field.

11. (a) $1s^6 2s^6 2p^7$

12. (c) $\Delta V = 4\pi r^2 \Delta r$

13. (b)

Chapter 14

1. (a)

2. (b)

3. (c)

4. (c)

5. (c) KE = E - PE = -8 MeV - (-40MeV)

6. (b) -8MeV + 15 MeV

7. ${}^1_0n$

8. (a) Mass of parent for beta plus decay must be greater than the mass of the daughter plus two electron masses. For electron capture, the mass of parent must only be greater than the mass of the daughter.

9. (b)

10. (b) When a nucleus decays by beta minus or beta plus, there is no change in the mass number A. When a nucleus decays by alpha decay the mass number decreases by four. Note that (237-209) = 28 = 4 x 7 while neither (237 - 208) nor (237 - 206) equals an integer times 4.

11. (a) $PE = -k_e e^2/r$ in joules. To convert to MeV divide by 1.6×10^{-13}.

12. (a) 3/2 kT = energy

Chapter 15

1. (c) Range = $c\Delta t = c(h/2\pi\Delta E)$, where ΔE equals the rest mass energy. The shorter the range, the less is the contribution to the nuclear force.

2. (c)

3. (c)

4. (a) $K^- + p \rightarrow K^+ + \Sigma^-$
S: $-1 + 0 = 1 - 2$

5. (c)

6. (b)

7. (a) $\Delta t = (6.6 \times 10^{-22}/60)$ seconds $= 1.1 \times 10^{-23}$ corresponds to a time for a strong interaction.

8. (c)

9. (c) Fox X^o, I_z must equal -1 to conserve isotopic spin. There must exist states with $I_z = 0$ and $I_z = 1$.

10. (c)

11. (d)

Chapters 1 through 3

1. (c) Net force must equal zero.

2. (c) This is a vector addition.

3.

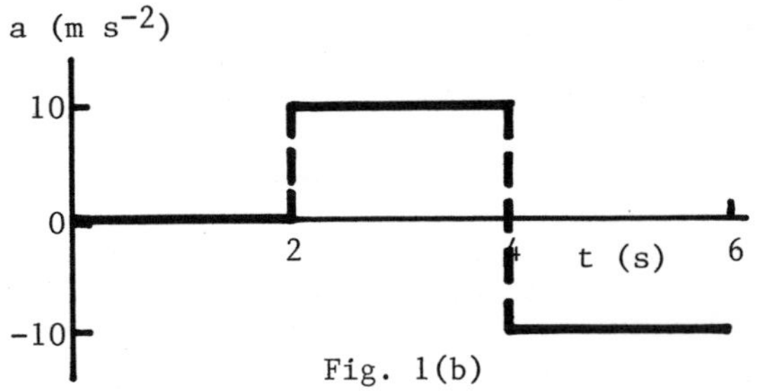

Fig. 1(b)

4. (e)

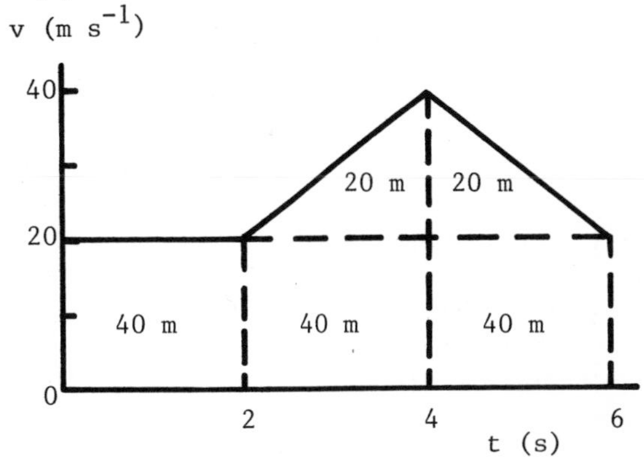

5. (b) First find the displacements and add them vectorially. The average velocity equals the resultant displacement divided by the total time.

6. (c) From the information about the vertical motion find the time t = 5s that the bullet is in the air. Then use the horizontal component of the velocity to find the distance moved horizontally.

7. (c) vertical velocity = $8 \text{ m s}^{-1} - (10\text{ms}^{-2})(5\text{s})$

8. (b) Net force = $(m_B + m_R)$ a

9. (b)

10. (c) First isolate the whole system and find the acceleration. Then isolate the 6kg object to find T_2.

Chapters 4 through 6

1. (c) The gravitational force on the object = mg = 100N. The net force equals zero for motion with constant velocity.

2. (d) Now net force = F' - mg = ma

3. (b) Don't forget the component of the weight down the plane.

4. (c) $v = 2\pi r/T$

5. (b) The seat must counterbalance the gravitational force and provide the centripetal acceleration into the center.

6. (d) Electric field between two parallel plates is constant.

7. (b)

8. (a) Don't forget that the radius = s/2.

9. (c) The particle experiences an acceleration in the electric field $a = qE/m$. The velocity acquired in the electric field $v = \sqrt{2as}$.

10. (a)

Chapters 7 and 8

1. (b) Work done = $\Delta PE + \Delta KE$

2. (a) $1/2\ kA^2 = 1/2\ kx^2 + 1/2\ mv^2$

3. (b) Use $1/2\ mv^2 = eV$ to find v.

4. (a) Kelvin temperature is proportional to the square of the velocity.

5. (c) Angular momentum is conserved. The potential energy increases with an increase in separation between the electron and the proton.

6. (b) The total current equals $(\xi_1 - \xi_2)$ divided by the total resistance of the circuit.

7. (c) The potential difference from a to c equals $(\xi_2 + ir_2)$

8. (b) $Pt = (V^2/R)t = m \times 4.18 \times 10^3 \times \Delta T$

9. (b)

10. (c)

11. (b) Don't forget that q = -6 cm.

12. (c) The first image is formed 25 cm to the right of the converging lens or 5 cm to the right of the diverging lens. Since this is a virtual object take p = -5 cm for the diverging lens.

For Chapters 9 and 10

1. (c) The frequency remains constant and $\lambda = v/f$

2. (b) In Fig. 1(b), the wave has moved $\lambda/4$ so that T = 1/5 s and $v = \lambda/T$.

3. (b) $a = -4\pi^2 y/T^2$. $a \neq -4\pi^2 s/T^2$

4. (c) $3/2\ \lambda = L$. $f = v/\lambda$.

5. (a) $1\lambda/4d = \sin\theta$

6. (a) Path difference = $\lambda/2$

7.

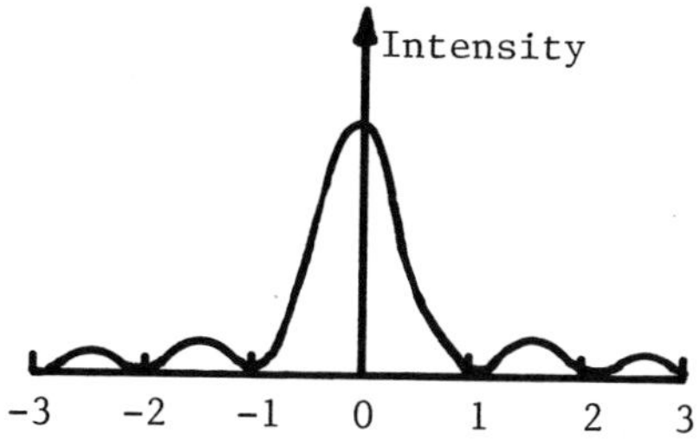

Fig. 4

8. (c) Note that $\Delta B = 0.6\ N(A-m)^{-1}$ into the page.

9. (a)

10. (b) $10V = \Delta\phi_B = \Delta(BA) = \Delta(B\ell x)$

Chapters 11 and 12

1. (b) This is a proper time.

2. (c)

3. (b)

4. (b) $u = (-3/5\ c + 4/5\ c)/(1-12/25)$

5. (c) $E = 12\ (1/\sqrt{1-(4/5)^2} + 1/\sqrt{1-(7/25)^2})$

6. (c)

7. (c)

8. (c)

9. (b)

10. (c)

11. (b)

12. (b)

13. (d)

14. (c)

Chapters 13 through 15

1. (c) $PE = -k_e e^2/2r = 1/2\ E$

2. (b) $13.6\ (Z-1)^2 eV(1-3/4) = \Delta E$
$\lambda = 1.24 \times 10^{-6}/\Delta E$

3. (c) $E_n = n^2 E_1$

4. (c) For sodium $E_n = -13.6\ Z_{eff}\ eV/n^2$

5. (b) $(238 - 206) = 32 = 8 \times 4$

6. (d)

7. (c) For electron capture $Q = (M_x - M_y)c^2$
For beta plus $Q = (M_x - M_y - 2\ m_e)c^2$

8. (c) Conserves strangeness.

9. (b) Time associated with weak decay since $\Delta S \neq 0$

10. (c)